Introduction to MATLAB®

Problem one

- go to curve fit app
- Select correct data
- Export to work space

Introduction to MATLAB®

Third Edition

Delores M. Etter
Southern Methodist University
Dallas, Texas

PEARSON

Upper Saddle River • Boston • Columbus
San Francisco • New York • Indianapolis • London
Toronto • Sydney • Singapore • Tokyo • Montreal
Dubai • Madrid • Hong Kong • Mexico City
Munich • Paris • Amsterdam • Cape Town

Vice President and Editorial Director, ECS: *Marcia Horton*
Executive Editor: *Holly Stark*
Editorial Assistant: *Carlin Heinle*
Director of Marketing: *Margaret Waples*
Marketing Manager: *Tim Galligan*
Marketing Assistant: *Jon Bryant*
Program Management Team Lead: *Scott Disanno*
Program Manager: *Clare Romeo*
Project Manager: *Priyadharshini Dhanagopal*
Senior Operations Specialist: *Nick Sklitsis*
Operations Specialist: *Linda Sager*
Permissions Project Manager: *Karen Sanatar*
Full-Service Project Management: *Pavithra Jayapaul, Jouve*
Printer/Binder: *Courier Kendallville*
Typeface: *10/12 ITC New Baskerville Std*

Library of Congress Cataloging-in-Publication Data
Etter, Delores M. (Delores Maria), 1947–
Introduction to MATLAB/Delores M. Etter, Southern Methodist University, Dallas, Texas.—Third edition.
pages cm
ISBN-13: 978-0-13-377001-8
ISBN-10: 0-13-377001-X
1. Engineering mathematics—Data processing. 2. MATLAB. 3. Numerical analysis—Data processing. I. Title.
TA345.E8724 2014
620.001'51—dc23

2013046421

10 9 8 7 6 5 4 3 2

PEARSON

ISBN-13: 978-0-13-377001-8
ISBN-10: 0-13-377001-X

Contents

About This Book

Engineers and scientists use computers to solve a variety of problems, ranging from the evaluation of a simple function to solving a system of equations. MATLAB® has become the **technical computing environment** of choice for many engineers and scientists because it is a single interactive system that incorporates **numeric computations, scientific visualization**, and **symbolic computation**.

Because the MATLAB computing environment is one that a new engineer is likely to encounter in a job, it is a good choice for an introduction to computing for engineers. This book is appropriate for use as an introductory engineering text or as a supplemental text in an advanced course. It is also useful as a professional reference.

This text was written to introduce engineering problem solving with the following objectives:

- Present a consistent **methodology for solving engineering problems**.
- Describe the exceptional **computational and visualization capabilities of MATLAB**.
- Illustrate the problem-solving process through a variety of **engineering examples and applications**.

To accomplish these objectives, Chapter 1 presents a five-step process that was developed by the author and is used to consistently solve engineering problems throughout the text. The rest of the chapters present the capabilities of MATLAB for solving engineering problems using specific examples from many different engineering disciplines.

TEXT ORGANIZATION

This book is designed for use in a variety of engineering and science course settings as a **primary text** for introductory students and as a **supplement for intermediate or advanced courses**. It is feasible to cover Chapters 1 through 8 in a one-semester course for a complete introduction to MATLAB's capabilities. If a briefer introduction to MATLAB is desired, we suggest that Chapters 1 through 4 be covered along with selected topics from Chapters 5 through 8.

PREREQUISITES

No prior experience with the computer is assumed. The mathematical background needed for Chapters 1 through 6 is **college algebra** and **trigonometry**. More advanced mathematics is needed for some of the material in later chapters.

PROBLEM-SOLVING METHODOLOGY

The emphasis on engineering and scientific problem solving is an important part of this text. Chapter 1 introduces a **five-step process for solving engineering problems** using the computer:

1. State the problem clearly.
2. Describe the input and output information.

3. Work a simple example by hand.
4. Develop an algorithm and convert it to MATLAB.
5. Test the solution with a variety of data.

To reinforce the development of problem-solving skills, each of these steps is identified every time a complete solution to an engineering problem is developed.

ENGINEERING AND SCIENTIFIC APPLICATIONS

Throughout the text, emphasis is placed on incorporating **real-world engineering and scientific examples** with solutions and usable code. Each chapter begins with a discussion of a significant engineering achievement. The chapter then includes examples related to this achievement. These examples include analysis of data from the following applications:

- temperatures from a sensor
- mass of air in a wind tunnel
- velocity and acceleration for unducted fan engine
- saturation vapor pressure for water at different temperatures
- Great Circle distances using GPS coordinates
- wind speeds from the Mount Washington Observatory
- wind speeds generated for a flight simulator
- ocean wave interaction
- performance quality scores
- mass calculations for a spacecraft
- current values for electrical circuits
- projectile range and impact
- interpolation using steam tables
- flow model for water in a culvert
- population models

VISUALIZATION

The visualization of the information related to a problem is a key advantage of using MATLAB for developing and understanding solutions. Therefore, it is important to learn to generate **plots** in a variety of formats to use when analyzing, interpreting, and evaluating data. We begin using plots with the first MATLAB program presented in Chapter 1 and continually expand **plotting capabilities** within the remaining chapters. Chapter 4 covers all the main types of plots and graphs.

SOFTWARE ENGINEERING CONCEPTS

Engineers and scientists are also expected to develop and implement **user-friendly** and **reusable** computer solutions. Therefore, learning software engineering techniques is crucial to successfully develop these computer solutions. Readability and documentation are stressed in the development of programs. Through MATLAB, users are able to write portable code that can be transferred from one computer platform to another. Additional topics that relate to software engineering issues are discussed in Chapter 1 and include the **software life cycle**, **maintenance**, and **software prototypes**.

PROBLEM SETS

Learning any new skill requires practice at a number of **different levels of difficulty**. Each chapter ends with a set of problems. These are problems that relate to a variety of engineering applications with the level of difficulty ranging from straightforward to longer assignments. **Engineering data sets** are included, for many of the problems, to use in testing.

STUDENT AIDS

Each chapter ends with a **Summary** that reviews the topics covered and includes a list of **Key Terms**. A **MATLAB Summary** lists all the special symbols, commands, and functions defined in the chapter. **Hints** are provided to help the student avoid some of the common errors.

WHAT'S NEW IN THIS EDITION?

- The discussions, screen captures, examples, and problem solutions have been updated to reflect MATLAB Version 8.2, R2013b.
- A discussion of the new Help browser is included along with screen captures to illustrate using this feature.
- The section on random number generation has been rewritten to reflect changes relative to the random number seed and to include the new function for generating random integers.
- The section on numerical integration has been rewritten to support the new integration function.
- Updated examples and discussion for current hardware and software are included throughout the text.
- Updated discussions and examples of importing and exporting data with other applications, such as Excel.

Acknowledgments

I want to acknowledge the outstanding work of the publishing team at Prentice Hall. My first MATLAB text was published in 1993, so some of us have worked together for many years. I would like to especially acknowledge the support of Marcia Horton, Holly Stark, Clare Romeo, Scott Disanno, and Greg Dulles. I would also like to express my gratitude to my husband, a mechanical/aerospace engineer, for his help in developing some of the engineering applications. Finally, I want to recognize the important contributions of the many students in my introductory courses for their feedback on the explanations, the examples, and the problems.

Delores M. Etter
Texas Instruments Distinguished Chair in Engineering Education
Professor, Department of Electrical Engineering
The Bobby B. Lyle School of Engineering
Southern Methodist University
Dallas, Texas

CHAPTER

1 An Introduction to Engineering Problem Solving

Objectives

After reading this chapter, you should be able to

- describe some important engineering achievements,
- understand the relationship of MATLAB with computer hardware and software, and
- describe a five-step process for solving engineering problems.

ENGINEERING ACHIEVEMENTS

Engineers solve real-world problems using scientific principles from disciplines that include computer science, mathematics, physics, biology, and chemistry. It is this variety of subjects and the challenge of real problems that have a positive impact on our world, which makes engineering so interesting and rewarding. For example, engineers are working to develop techniques to provide access to clean water to people around the world. Engineers are working to make solar energy more economical so that it can give more than the 1 percent of our energy that it provides today. Engineers work to reduce pollution through developing ways to capture and store excess carbon dioxide in our manufacturing plants. Engineers restore and improve our urban and transportation infrastructure. At the beginning of each chapter, we will present a short discussion on a significant engineering achievement, and in that chapter, we will solve small problems related to that application.

1.1 ENGINEERING ENVIRONMENT

Engineers work in an environment that requires a **strong technical background**, and the computer will be the primary computational tool of most engineers. The focus of this text is to teach you the fundamentals of one of the most widely used engineering tools—MATLAB. However, engineers in the twenty-first century must also have many nontechnical skills and capabilities. The computer is also useful in developing additional nontechnical abilities.

Engineers need **strong communication skills** both for oral presentations and for the preparation of written material. Computers provide the software to assist in writing outlines and developing materials, such as graphs, for presentations and technical reports.

The **design/process/manufacture path**, which consists of taking an idea from a concept to a product, is one that engineers must understand firsthand. Computers are used in every step of this process, from design analysis, machine control, robotic assembly, quality assurance to market analysis.

Engineering teams today are **interdisciplinary teams**. Learning to interact in teams and to develop organizational structures for effective team communication is important for engineers. You are likely to be part of a diverse engineering team in which members are located around the globe, and thus, there are many additional challenges for teams that are not geographically located.

The engineering world is a **global** one. To be effective, you need to understand different cultures, political systems, and business environments. Courses in these topics and in foreign languages help provide some understanding, but exchange programs with international experiences provide invaluable knowledge in developing a broader understanding of the world.

Engineers are **problem solvers**, but problems are not always formulated carefully in the real world. An engineer must be able to extract a problem statement from a problem discussion and then determine the important issues. This involves not only developing order, but also learning to correlate chaos. It means not only analyzing the data, but also synthesizing a solution using many pieces of information. The integration of ideas can be as important as the decomposition of the problem into manageable pieces. A problem solution may involve not only abstract thinking about the problem, but also experimental learning from the problem environment.

Problem solutions must also be considered in their **societal context**. Environmental concerns should be addressed as alternative solutions to problems are being considered. Engineers must also be conscious of ethical issues in providing test results, quality verifications, and design limitations. Ethical issues are never easy to resolve, and some of the exciting new technological achievements bring ethical issues with them.

The material presented in this text is only one step in building the knowledge, confidence, and understanding needed by engineers today. We begin the process with a brief discussion of computing systems and an introduction to a problem-solving methodology that will be used throughout this text as we use MATLAB to solve engineering problems.

1.2 ENGINEERING COMPUTING AND MATLAB

Before we begin discussing MATLAB, a brief discussion on computing is useful, especially for those who have not had lots of experience with computers. A computer is a machine that is designed to perform operations that are specified with a set of instructions called a program. Computer ***hardware*** refers to the computer equipment, such as a notebook computer, a thumb drive, a keyboard, a flat-screen monitor, or a printer. Computer ***software*** refers to the programs that describe the steps we want the computer to perform. This can be software that we have written, or it can be programs that we download or purchase, such as computer games. Our computer hardware/software can be self-contained, as in a notebook computer. A computer can also access both hardware and software through a computer network, and through access to the Internet. In fact, cloud computing provides access to hardware, software, and large data sets through remote networks.

1.2.1 Computer Hardware

All computers have a common internal organization as shown in Figure 1.1. The processor is the part of the computer that controls all the other parts. It accepts input values (from a device such as a keyboard or a data file) and stores them in memory. It also interprets the instructions in a computer program. If we want to add two values, the processor will retrieve the values from memory and send them to the arithmetic logic unit (ALU). The ALU performs the addition, and the processor then stores the result in memory. The processing unit and the ALU use internal memory composed of read-only memory (ROM) and random access memory (RAM); data can also be stored in external storage devices such as external drives or thumb drives. The processor and the ALU together are called the ***central processing unit*** (CPU). A ***microprocessor*** is a CPU that is contained in a single integrated-circuit chip, which contains millions of components in an area much smaller than a postage stamp.

Many inexpensive printers today use ink-jet technology to print both color copies and black-and-white copies. We can also store information on a variety of digital memory devices, including CDs and DVDs. A printed copy of information is called a ***hard copy***, and a digital copy of information is called an ***electronic copy*** or a ***soft copy***. Many printers today can also perform other functions such as copying, faxing, and scanning.

Computers come in all sizes, shapes, and forms. In fact, most of our phones today contain CPUs and store programs that they can execute. Smartphones also contain a graphics processing unit, a significant amount of RAM, and are trending to multicore (or multiprocessor), low-power CPUs. Many homes today have

Figure 1.1
Internal organization of a computer.

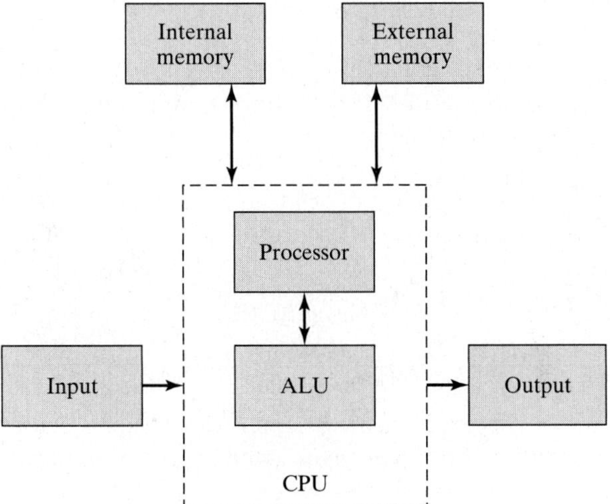

personal computers that are used for a variety of applications, including e-mail, financial budgeting, and games; these computers are typically desktop computers with separate monitors and keyboards. Notebook computers contain all their hardware in a small footprint, and thus become very convenient. For some people, tablet computers (such as the iPad) and smartphones are even replacing the use of the desktop and notebook computers.

1.2.2 Computer Software

Computer software contains the instructions or commands that we want the computer to perform. There are several important categories of software, including operating systems, software tools (MATLAB is a software tool), and language compilers. Figure 1.2 illustrates the interaction among these categories of software and the computer hardware. We now discuss each of these software categories in more detail.

Operating Systems

Some software, such as an operating system, typically comes with the computer hardware when it is purchased. The ***operating system*** provides an interface between you (the user) and the hardware by providing a convenient and efficient environment in which you can select and execute the software application on your system. The component of the operating system that manages the interface between the hardware and software applications is called a ***kernel***. Examples of desktop operating systems include Windows, Mac OS, Unix, and Linux. Operating systems for smartphones include Android (a Linux variant) and iOS (a Unix variant).

Operating systems also contain a group of programs called ***utilities*** that allow you to perform functions such as printing files, copying files from one folder to another, and listing the files in a folder. Most operating systems today simplify using these utilities through icons and menus.

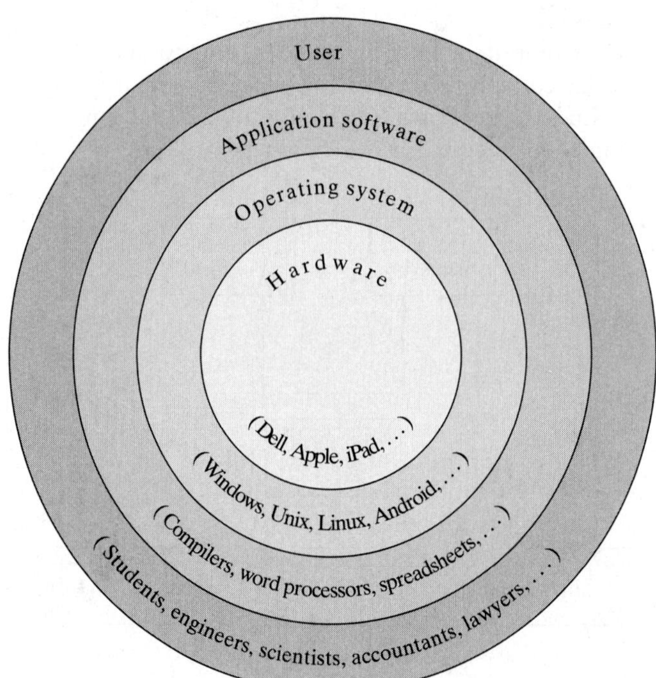

Figure 1.2
Interactions between software and hardware.

Software Tools

Software tools are programs that have been written to perform common operations. For example, **word processors** like Microsoft Word are programs that allow you to enter and format text. They allow you to download information from the Internet into the file, and allow you to enter mathematical equations. They also can check your grammar and spelling. Most word processors also allow you to produce documents that have figures, images, and can print in two columns. These capabilities allow you to perform desktop publishing from a notebook computer.

Spreadsheet programs like Excel are software tools that allow you to easily work with data that can be displayed in a grid of rows and columns. Spreadsheets were initially developed to be used for financial and accounting applications, but many science and engineering problems can be easily solved with spreadsheets. Most spreadsheet packages include plotting capabilities, so they are especially useful in analyzing and displaying information in charts. Database management tools allow you to analyze and "mine" information from large data sets.

Another important category of software tools is **mathematical computation tools**. This category includes MATLAB and Mathematica. Not only do these tools have very powerful mathematical commands, but they are also graphics tools that provide extensive capabilities for generating graphs. This combination of computational and visualization power make them particularly useful tools for engineers.

If an engineering problem can be solved using a software tool, it is usually more efficient to use the software tool than to write a program in a computer language. The distinction between a software tool and a computer language is becoming less clear as some of the more powerful software tools include their own language in addition to having specialized operations. (MATLAB is both a software tool and a programming language.)

Computer Languages

Computer languages can be described in terms of generations. The first generation of computer languages is machine languages. ***Machine languages*** are tied closely to the design of the computer hardware, and are often written in binary strings consisting of 0s and 1s. Therefore, machine language is also called binary language.

An ***assembly language*** is also unique to a specific computer design, but its instructions are written in symbolic statements instead of binary. Assembly languages usually do not have many statements; thus, writing programs in assembly language can be tedious. In addition, to use an assembly language you must also know information that relates to the specific hardware. Instrumentation that contains microprocessors often requires that the programs operate very fast; thus, the programs are called real-time programs. These real-time programs are usually written in assembly language to take advantage of the specific computer hardware in order to perform the steps faster. Assembly languages are second generation languages.

Third generation languages use English-like commands. These languages include C, C++, C#, and Java. Writing programs in a ***high-level language*** is certainly easier than writing programs in machine language or in assembly language. However, a high-level language contains a large number of commands and an extensive set of ***syntax*** (or grammar) rules for using these commands.

MATLAB is considered a fourth generation programming language because of its powerful commands, user interfaces, and its ability to interface to other languages. Higher-level languages are still primarily in research phases and tend to be domain specific.

1.2.3 Executing a Computer Program

A program written in a high-level language such as C must be translated into machine language before the instructions can be executed by the computer. A special program called a ***compiler*** is used to perform this translation. Thus, in order to write and execute C programs, we must have a C compiler. The C compilers are available as separate software packages for use with specific operating systems.

If any errors (often called ***bugs***) are detected by the compiler during compilation, corresponding error messages are printed. We must correct our program statements and then perform the compilation step again. The errors identified during this stage are called ***compiler errors*** or compile-time errors. For example, if we want to divide the value stored in a variable called sum by 3, the correct expression in C is sum/3; if we incorrectly write the expression using the backslash, as in sum\3, we will get a compiler error. The process of compiling, correcting statements (or ***debugging***), and recompiling is often repeated several times before the program compiles without compiler errors. When there are no compiler errors, the compiler generates a program in machine language that performs the steps specified by the original C program. The original C program is referred to as the source program, and the machine language version is called an object program. Thus, the source program and the object program specify the same steps; but the source program is written in a high-level language, and the object program is specified in machine language.

Once the program has compiled correctly, additional steps are necessary to prepare the object program for execution. This preparation involves linking other machine language statements to the object program and then loading the program into memory. After this linking/loading, the program steps are executed by the computer. New errors called execution errors, run-time errors, or ***logic errors*** may be identified in this stage; they are also called program bugs. Execution errors often cause the termination of a program. For example, the program statements may attempt to perform a division by zero, which generates an execution error. Some execution errors do not stop the program from executing, but they cause incorrect results to be computed. These types of errors can be caused by programmer errors in determining the correct steps in the solutions and by errors in the data processed by the program. When execution errors occur due to errors in the program statements, we must correct the errors in the source program and then begin again with the compilation step. Even when a program appears to execute properly, we must check the answers carefully to be sure that they are correct. The computer will perform the steps precisely as we specify, and if we specify the wrong steps, the computer will execute these wrong (but syntactically legal) steps and thus present us with an answer that is incorrect.

The process of compilation, linking/loading, and execution is outlined in Figure 1.3. The process of converting an assembly language program to binary is performed by an assembler program, and the corresponding processes are called assembly, linking/loading, and execution.

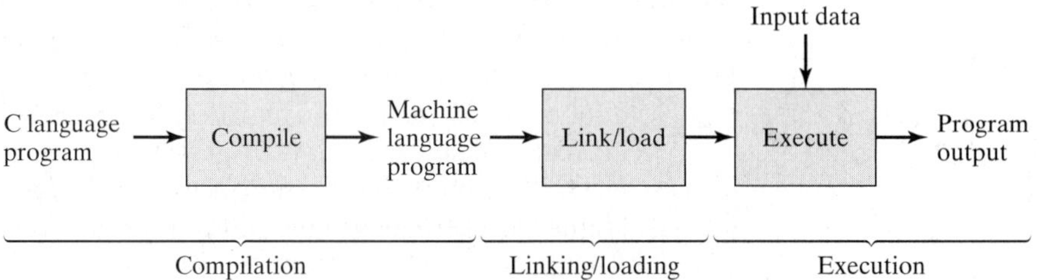

Figure 1.3
Program compilation/loading, linking, and execution.

1.2.4 Software Life Cycle

The cost of a computer solution to a problem can be estimated in terms of the cost of the hardware and the cost of the software. The majority of the cost in a computer solution today is in the cost of the software, and thus, a great deal of attention has been given to understanding the development of a software solution.

The development of a software project generally follows definite steps or cycles, which are collectively called the ***software life cycle***. These steps typically include project definition, detailed specification, coding and modular testing, integrated testing, and maintenance. (These steps will be explained in more detail in later chapters.) Software maintenance is a significant part of the cost of a software system. This maintenance includes adding enhancements to the software, fixing errors identified as the software is used, and adapting the software to work with new hardware and software. The ease of providing maintenance is directly related to the original definition and specification of the solution because these steps lay the foundation for the rest of the project. The problem-solving process that we present in the next section emphasizes the need to define and specify the solution carefully before beginning to code or test it.

One of the techniques that has been successful in reducing the cost of software development both in time and cost is the development of ***software prototypes***. Instead of waiting until the software system is developed and then letting the users work with it, a prototype of the system is developed early in the life cycle. This prototype does not have all the functions required of the final software, but it allows the user to use it early in the life cycle, and to make desired modifications to the specifications. Making changes earlier in the life cycle is both cost- and time-effective. It is not uncommon for a software prototype to be developed in MATLAB, and then for the final system to be developed in another language.

As an engineer, it is very likely that you will need to modify or add additional capabilities to existing software that has been developed using a software tool or a high-level language. These modifications will be much simpler if the existing software is well-structured and readable and if the documentation that accompanies the software is up-to-date and clearly written. For these reasons, we stress developing good habits that make programs more readable and self-documenting.

1.3 AN ENGINEERING PROBLEM-SOLVING METHODOLOGY

Problem solving is a key part not only of engineering courses, but also of courses in computer science, mathematics, physics, and chemistry. Therefore, it is important to have a consistent approach to solving problems. It is also helpful if the approach is general enough to work for all these different areas, so that we do not have to learn one technique for solving mathematics problems, a different technique for solving physics problems, and so on. The problem-solving process that we present works for engineering problems and can be tailored to solve problems in other areas as well. However, it does assume that we are using a computer to help solve the problem.

The process, or methodology, for problem solving that we will use throughout this text has five steps:

1. State the problem clearly.
2. Describe the input and output information.
3. Work the problem by hand (or with a calculator) for a simple set of data.
4. Develop a MATLAB solution.
5. Test the solution with a variety of data.

We now discuss each of these steps using data collected from a physics laboratory experiment as an example.

EXAMPLE 1.1

TEMPERATURE ANALYSIS AND PLOT

Assume that we have collected a set of temperatures from a sensor on a piece of equipment that is being used in an experiment. The temperature measurements shown in Table 1.1 are taken every 30 seconds, for 5 minutes, during the experiment. We want to compute the average temperature, and we also want to plot the temperature values.

Table 1.1 Experimental Temperature Data

Time, minutes	Temperature, F
0.0	105
0.5	126
1.0	119
1.5	129
2.0	132
2.5	128
3.0	131
3.5	135
4.0	136
4.5	132
5.0	137

SOLUTION

1. Problem Statement

The first step is to state the problem clearly. It is extremely important to give a clear, concise statement of the problem, in order to avoid any misunderstandings. For this example, the statement of the problem is as follows:

Compute the average of a set of temperatures. Then plot the time and temperature values.

2. Input/Output Description

The second step is to describe carefully the information that is given to solve the problem and then to identify the values to be computed. These items represent the input and the output for the problem and collectively can be called input/output, or I/O. For many problems, it is useful to create a diagram that shows the input and output. At this point, the program is called an abstraction because we are not defining the steps to determine the output; instead, we are only showing the information that is used to compute the output. The I/O diagram for this example is as follows:

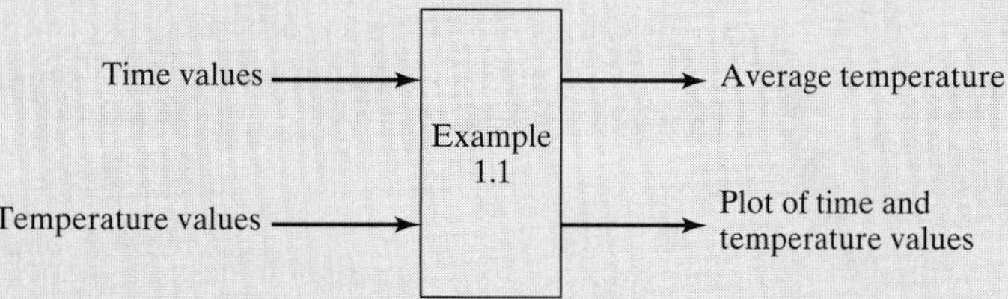

3. Hand Example

The third step is to work the problem by hand or with a calculator, using a simple set of data. This step is very important and should not be skipped, even for simple problems. This is the step in which you work out the details of the solution to the problem. If you cannot take a simple set of numbers and compute the output (either by hand or with a calculator), you are not ready to move on to the next step. You should reread the problem and perhaps consult reference material. For this problem, the only calculation is computing the average, or mean value, of a set of temperature values. Assume that we use the first three sets of data for the hand example. By hand, we compute the average to be $(105 + 126 + 119)/3$, or 116.6667.

4. MATLAB Solution

Once you can work the problem for a simple set of data, you are ready to develop an ***algorithm***, which is a step-by-step outline of the solution to the problem. For simple problems such as this one, the algorithm can be written immediately using MATLAB commands. For more complicated problems, it may be necessary to write an outline of the steps and then decompose the steps into smaller steps that can be translated into MATLAB commands. One of the strengths of MATLAB is that its commands match very closely to the steps that we use to solve engineering problems. Thus, the process of determining the steps to solve the problem also determines the MATLAB commands. At this point, we know that you do not yet understand the MATLAB commands. However, we present the solution so you can observe that the MATLAB steps match closely to the solution steps from the hand example:

```
%------------------------------------------------------------
%  Example 1_1 This program computes the average
%  temperature and plots the temperature data.
%
time = [0.0,0.5,1.0];
temps = [105,126,119];
average = mean(temps)
plot(time,temps),title('Temperature Measurements'),
    xlabel('Time, minutes'),
    ylabel('Temperature, degrees F'),grid
%------------------------------------------------------------
```

The words that follow percent signs are comments to help us in reading the MATLAB statements. If a MATLAB statement assigns or computes a value, it will also print the value on the screen if the statement does not end in a semicolon. Thus, the values of **time** and **temps** will not be printed, because the statements that assign

(*continued*)

them values end with semicolons. The value of the average will be computed and printed on the screen, because the statement that computes it does not end with a semicolon. Finally, a plot of the time and temperature data will be generated.

5 Testing

The final step in our problem-solving process is testing the solution. We should first test the solution with the data from the hand example, because we have already computed the solution to it. When the previous statements are executed, the computer displays the following output:

```
average =
         116.6667
```

A plot of the data points is also shown on the screen. Because the value of the average computed by the program matches the value from the hand example, we now replace the data from the hand example with the data from the physics experiment using these replacement statements:

```
time = [0.0,0.5,1.0,1.5,2.0,2.5,3.0,3.5,4.0,4.5,5.0];
temps = [105,126,119,129,132,128,131,135,136,132,137];
```

When the commands in the program are executed with the complete set of data, the computer displays the following output:

```
average =
         128.1818
```

The plot in Figure 1.4 is also shown on the screen.

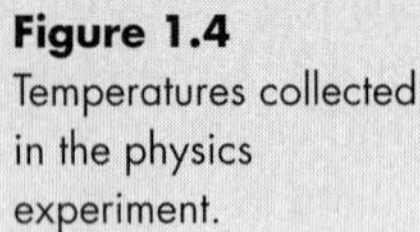

Figure 1.4
Temperatures collected in the physics experiment.

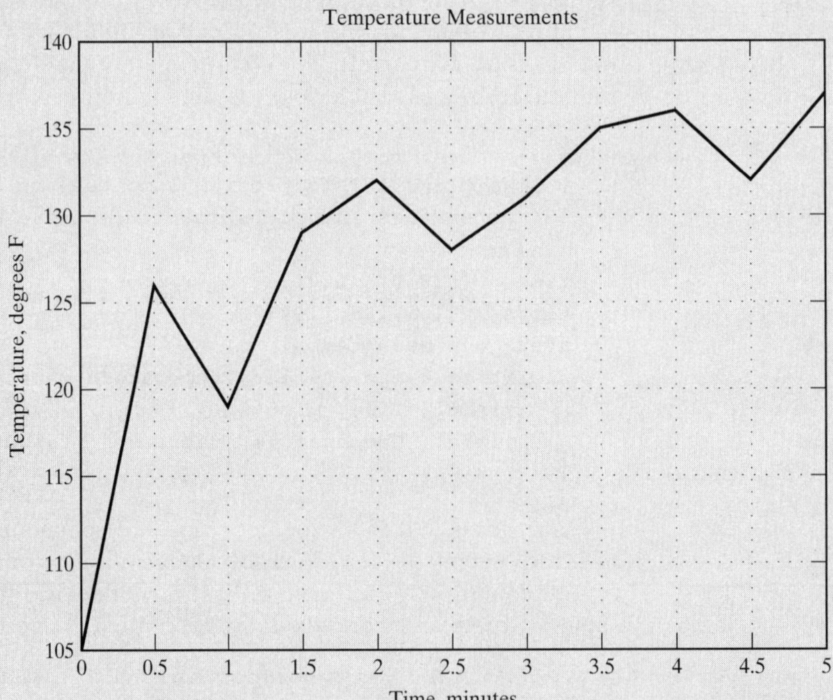

SUMMARY

Engineers solve real-world problems, and most of these solutions require the use of computing to help develop good solutions. We presented a summary of the components of a computer system from the computer hardware to computer software, and reviewed the different types of computer languages and software tools that help us develop problem solutions. We then introduced a five-step problem-solving process that we have used throughout this text. These five steps are:

1. State the problem clearly.
2. Describe the input and output information.
3. Work the problem by hand (or with a calculator) for a simple set of data.
4. Develop a MATLAB solution.
5. Test the solution with a variety of data.

KEY TERMS

algorithm
assembly language
bugs
central processing unit
compiler
compiler errors
debugging
electronic copy
hard copy
hardware
high-level languages
kernel
logic errors
machine languages
microprocessor
operating system
soft copy
software
software life cycle
software prototypes
syntax
utilities

PROBLEMS

The solutions to these problems are at the end of this chapter.

TRUE-FALSE PROBLEMS

Indicate whether the following statements are true (T) or false (F):

1.	A CPU consists of an ALU, memory, and a processor.	T	F
2.	Linking/loading is the step that prepares the object program for execution.	T	F
3.	An algorithm describes the problem solution step-by-step, while a computer program solves the problem in one step.	T	F
4.	A computer program is the implementation of an algorithm.	T	F
5.	A utility program converts a high-level language to binary.	T	F
6.	A microprocessor is a processor that is very small.	T	F
7.	Data can be communicated between internal memory and external memory through an ALU.	T	F
8.	Spreadsheets are useful to manipulate objects graphically.	T	F

9. MATLAB is a commonly-used software tool. T F

10. A word processor allows you to enter and edit text. T F

11. MATLAB is very powerful both in computation and in visualization. T F

12. To correct logic errors we repeat only the execution step. T F

13. The compilation step identifies all the bugs in the program. T F

14. An algorithm gives the steps used to solve a problem. T F

15. A computer program is a set of instructions to solve a problem. T F

16. A program is completely tested if it works for one set of data. T F

17. A thumb drive is an example of a soft copy. T F

18. C++ is an assembly language. T F

19. Software maintenance is an insignificant part of the cost of today's software systems. T F

20. Programs written in machine language can be represented in binary. T F

MULTIPLE-CHOICE PROBLEMS

Circle the letter for the best answer to complete each statement:

21. Instructions and data are stored in

(a) the arithmetic logic unit (ALU).
(b) the control unit (processor).
(c) the central processing unit (CPU).
(d) the memory.
(e) the keyboard.

22. An operating system is

(a) the software that is designed by users.
(b) a convenient and efficient interface between the user and the hardware.
(c) the set of utilities that allows us to perform common operations.
(d) a set of software tools.

23. Source code is

(a) the result of compiler operations.
(b) the process of getting information from the processor.
(c) the set of instructions in a computer language that solves a specific problem.
(d) the data stored in the computer memory.
(e) the values entered through the keyboard.

24. Object code is

(a) the result of compiler operations on the source code.
(b) the process of obtaining information from the processor.
(c) a computer program.
(d) a process involving the listing of commands required to solve a specific problem.
(e) the result of the linking/loading process.

25. An algorithm refers to

(a) a step-by-step solution to solve a specific problem.
(b) a collection of instructions that the computer can understand.

(c) a code that allows us to type in text materials.
(d) stepwise refinement.
(e) a set of math equations to derive the solution to a problem.

26. A hard copy is

(a) the information stored on a hard disk.
(b) the information printed out on paper.
(c) the information shown on the screen.
(d) a computer program.
(e) all of the above.

27. Computer hardware consists of

(a) a keyboard.
(b) a printer.
(c) internal memory.
(d) a terminal screen.
(e) all of the above.

28. An example of software is

(a) a printer.
(b) a screen.
(c) a computer code.
(d) the memory.
(e) all of the above.

29. High-level languages are

(a) good for real-time programming.
(b) the second generation of computer languages.
(c) written in binary.
(d) written in English-like words.

30. The difference between the source program and the object program is

(a) the source program possibly contains some bugs, and the object program does not contain any bugs.
(b) the source program is the original code, and the object program is a modified code.
(c) the source program is specified in a high-level language, and the object program is specified in machine language.
(d) the object program is also a source program.
(e) the source program can be executed, and the object program cannot be executed.

31. The place to start when solving a problem is

(a) to develop an algorithm.
(b) to write the program.
(c) to compile the source program.
(d) to link to the object program.

32. A hand example means

(a) doing arithmetic on your hands.
(b) working out the details of the problem solution using a simple set of data.
(c) outlining a solution to a problem.
(d) expanding the outline of a solution.
(e) testing the algorithm step-by-step with a calculator.

33. A computer program is

(a) a collection of components containing the input and output devices.
(b) a list of instructions needed to solve a problem.
(c) a set of instructions to be performed by the computer and written in a language that a computer can understand.
(d) a step-by-step procedure for solving a problem.
(e) an outline that decomposes the problem into simpler steps.

MATCHING PROBLEMS

Select the correct term for each of the following definitions from this list:

algorithm	natural languages
arithmetic logic unit (ALU)	network
central processing unit (CPU)	operating systems
compilation	output devices
debugging	program
grammar	software life cycle
hardware	software maintenance
input devices	spreadsheet
logic errors	syntax
machine language	system software
memory	utility
microprocessor	word processor

34. A set of instructions that tells a computer what to do
35. The machinery that is part of the computer
36. The brain of the computer
37. Devices used to show the results of programs
38. Compilers and other programs that help run the computer
39. The steps to solve a problem
40. The process that converts a C program into machine language
41. A software tool designed to work with data stored in a grid or a table
42. The rules that define the punctuation and words that can be used in a program
43. The interface between the user and the hardware
44. The part of a computer that performs the mathematical computations
45. The process of removing errors from a program
46. Errors discovered during the execution of a program
47. The programs help you print files and copy files
48. A central processing unit contained in a single integrated-circuit chip
49. A software tool used to enter text
50. The representation of a program in binary

SOLUTIONS:

1. F
2. T
3. F
4. T
5. F
6. F
7. F
8. F
9. T
10. T
11. T
12. F
13. F
14. T
15. T
16. F
17. F
18. F
19. F
20. T
21. (d)
22. (b)
23. (c)
24. (a)
25. (a)
26. (b)
27. (e)
28. (c)
29. (d)
30. (c)
31. (a)
32. (b)
33. (c)
34. program
35. hardware
36. CPU
37. output devices
38. system software
39. algorithm
40. compilation
41. spreadsheet
42. syntax or grammar
43. operating system
44. ALU
45. debugging
46. logic errors
47. utilities
48. microprocessor
49. word processor
50. machine language

CHAPTER

2 Getting Started with MATLAB

Objectives

After reading this chapter, you should be able to

- understand the MATLAB screen layout, windows, and interactive environments,
- initialize and use scalars, vectors, and matrices in computations,
- write simple programs using MATLAB, and
- create and use script M-files.

ENGINEERING ACHIEVEMENT: WIND TUNNELS

Wind tunnels are test chambers built to generate precise wind speeds. Accurate scale models of new aircraft and missiles can be mounted on force-measuring supports in the test chamber, and then measurements of the forces acting on the models can be made at many different wind speeds and angles of the models relative to the wind direction. Some wind tunnels can operate at hypersonic velocities, generating wind speeds of thousands of miles per hour. The sizes of wind tunnel test sections vary from a few inches across to sizes large enough to accommodate a fighter jet. At the completion of a wind tunnel test series, many sets of data have been collected that can be used to determine the lift, drag, and other aerodynamic performance characteristics of a new aircraft at its various operating speeds and positions. Wind tunnels are also used to test the performance of sports equipment like composite skis, snowboards, bicycles, and racing cars. In this chapter, we give examples of using MATLAB to analyze wind tunnel results.

2.1 INTRODUCTION TO MATLAB AND MATLAB WINDOWS

MATLAB is one of a number of commercially available, sophisticated mathematical computation tools, such as Mathematica and MathCad. Despite what their proponents may claim, none of these tools is "the best." They all have strengths and weaknesses.

Each will allow you to perform basic mathematical computations, but they differ in the ways that they handle symbolic calculations and more complicated mathematical processes. MATLAB excels at computations involving matrices. In fact, its name, **MATLAB**, is short for **Mat**rix **Lab**oratory. At a very basic level, you can think of these programs as sophisticated, computer-based calculators. They can perform the same functions as your scientific calculator, but they can also do much more. In many engineering programs, students are learning to use mathematical computational tools like MATLAB, in addition to also learning a high-level language such as JAVA, C, or C++. This then gives you the option of choosing the right tool or language for the problem that you are solving.

Today's MATLAB has capabilities far beyond the original MATLAB and is an interactive system and programming language for general scientific and technical computation. Because MATLAB commands are similar to the way that we express engineering steps in mathematics, writing computer solutions in MATLAB can be much quicker than writing solutions in a high-level language. It is important to understand when to use a computational program such as MATLAB and when to use a general purpose, high-level programming language. MATLAB excels at numerical calculations, especially matrix calculations, and graphics. Usually, high-level programs do not offer easy access to graphing. The primary area of overlap between MATLAB and high-level programs is in "number crunching"—programs that require repetitive calculations or processing of large quantities of data. Both MATLAB and high-level languages are good at processing numbers. It is usually easier to write a "number crunching" program in MATLAB, but it usually executes faster in C or C++. The one exception to this rule is with matrices. Because MATLAB is optimized for matrices, if a problem can be formulated with a matrix solution, MATLAB executes substantially faster than a similar program in a high-level language.

HINT

A number of examples are presented in this text. We encourage you to type the example problems into MATLAB as you read the book, and observe the results. You may think that some of the examples are too simple to type in yourself—that just reading the material is sufficient. However, you will remember the material much better if you both read it and type it.

To begin MATLAB, use your mouse to click on the MATLAB icon on the desktop or choose it from the list of Applications on your computer. To exit MATLAB, use the close icon (x) from the upper right-hand corner of the screen for a PC or use the red circle at the upper left-hand corner of the screen for an Apple computer. These are essentially the only differences that you will see between PC's and Apple computers. The MATLAB screens and output will be the same. You should see the MATLAB **prompt >>** (or **EDU** >> if you are using the Student Edition) in the middle of the screen which tells you that MATLAB is waiting for you to enter a command. To exit MATLAB, type **`quit`** or **`exit`** at the MATLAB prompt, or select the close icon (x) from the top of the screen.

MATLAB uses display windows. The default view shown in Figure 2.1 includes a large command window in the center, the current folder window on the left, and

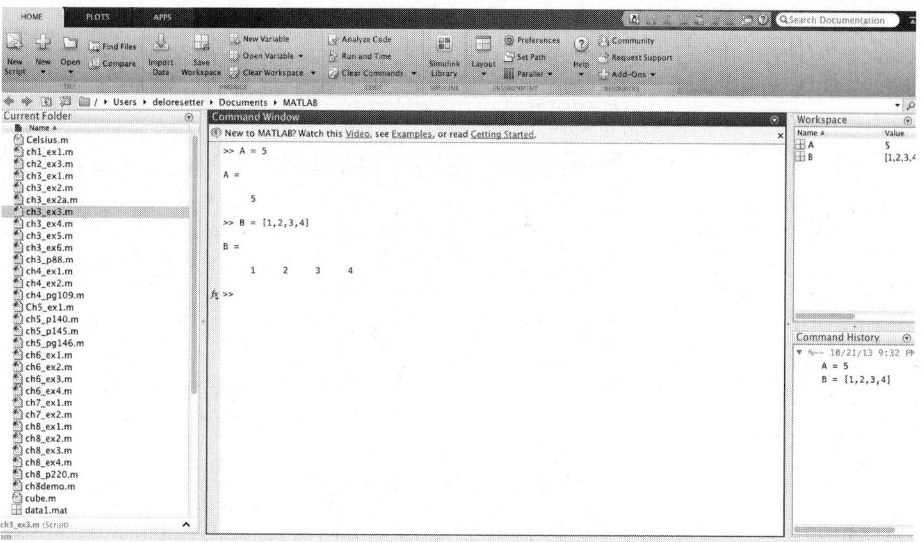

Figure 2.1
MATLAB opening window.

the workspace and command history windows on the right. In addition, document windows, graphics windows, and editing windows will automatically open when needed.

2.1.1 Command Window

You can use MATLAB in two basic modes. The **command window** offers an environment similar to a note pad. Using the command window allows you to save the values you calculate, but not the commands used to generate those values. If you want to save the command sequence, you will need to use the editing window to create a script or an **M-file**. (M-files are presented in Section 2.4.) Both approaches are valuable. In this chapter, we will concentrate on using the command window so that you become comfortable with it. In the later chapters, we will use M-files to store programs that we develop with MATLAB.

You can perform calculations in the command window in a manner very similar to the way you perform calculations on a scientific calculator. Most of the syntax is even the same. For example, to compute the value of 5 squared, type the command

```
5^2
```

The answer will be displayed in the following manner:

```
ans =
       25
```

Or, to use a trigonometric function to find the value of $\cos(\pi)$, type

```
cos(pi)
```

which results in the following output:

```
ans =
       -1
```

We will cover MATLAB functions in Chapter 3.

HINT

You may find it frustrating to discover that when you make a mistake, you cannot just overwrite your command on the same line after you have executed it. This is because the command window is creating a list of all the commands you have entered. Instead, you can enter the correct command on a new line, and then execute your new version. MATLAB also offers another way to make it easier to update commands. You can use the arrow keys, usually located on the right-hand side of your keyboard. The **up arrow**, ↑, allows you to move through the list of commands you have executed. Once you find the appropriate command, you can edit it, and then execute your new version. This can be a real time-saver.

2.1.2 Command History Window

The **command history window** records the commands you issued in the command window. When you exit MATLAB, or when you issue the **clc** command to clear all commands, the command window is cleared. However, the command history window retains a list of all of your commands. You may clear the command history using the edit menu if you need to. If you work on a public computer, as a security precaution, MATLAB's defaults may be set to clear the history when you exit MATLAB. If you entered the example commands above, notice that they are repeated in the command history window. This window is valuable for a number of reasons. It allows you to review previous MATLAB sessions, and it can be used to transfer commands to the command window. For example, in the command window, type

```
clc
```

This should clear the command window, but leave the data in the command history window intact. You can transfer any command from the command history window to the command window by double clicking (which also executes the command) or by clicking and dragging the line of code into the command window. Try double clicking

```
cos(pi)
```

which should return

```
ans =
      -1
```

Click and drag

```
5^2
```

from the command history window into the command window. The command will not execute until you press the enter key, and then you'll get the following result:

```
ans =
      25
```

You will find the command history useful as you perform more and more complicated calculations in the command window.

2.1.3 Workspace Window

The **workspace window** keeps track of the variables you have defined as you execute commands in the command window. As you do the examples, the workspace window should just show one variable, **ans**, and it should also tell us that it has a value of 25, as shown below:

Name	Value
ans	25

The workspace window can be cleared using the command **clear** or by using the edit menu.

Set the workspace window to show more about this variable by selecting Workspace actions in the Workspace header bar. Then select Choose Columns. Check **size** and **bytes**, in addition to **name**, **value**, and **class**. Your workspace window should now display:

Name	Value	Size	Bytes	Class
ans	25	1 × 1	8	double

The yellow grid-like symbol indicates the variable **ans** is an array. The size, 1 × 1, tells us that it is a single value (one row by one column) and therefore a scalar. The array uses 8 bytes of memory. MATLAB was written in C, and the class designation tells us that in the C language **ans** is a double precision, floating point array. For our needs it is enough to know that the variable **ans** can store a floating point number (one with a decimal point). MATLAB considers every number you enter to be a floating point number, whether you put a decimal in the number or not.

You can define additional variables in the command window and they will be listed in the workspace window. For example, type

```
A = 5
```

which returns

```
A =
     5
```

Notice that the variable A has been added to the workspace window, which lists variables in alphabetical order. Variables beginning with capital letters are listed first, followed by variables starting with lowercase letters:

Name	Value	Size	Bytes	Class
A	5	1 × 1	8	double
ans	25	1 × 1	8	double

Entering matrices into MATLAB is not discussed in detail in this section. However, you can enter a simple one-dimensional matrix by typing

```
B = [1, 2, 3, 4]
```

which returns

```
B =
    1   2   3   4
```

The commas are optional. You would get the same result with

```
B = [1 2 3 4]
```

Notice that the variable B has been added to the workspace window and that it is a 1×4 array:

Name	Value	Size	Bytes	Class
A	5	1 × 1	8	double
B	[1 2 3 4]	1 × 4	32	double
ans	25	1 × 1	8	double

We define two-dimensional matrices in a similar fashion. Semicolons are used to separate rows. For example,

```
C = [1,2,3,4; 10,20,30,40; 5,10,15,20]
```

displays

```
C =
     1    2    3    4
    10   20   30   40
     5   10   15   20
```

Notice that C appears in the workspace window as a 3×4 matrix. You can recall the values for any variable by just typing in the variable name. For example, entering

```
A
```

displays

```
A =
     5
```

The information that is shown in the workspace window is the following:

Name	Value	Size	Bytes	Class
A	5	1 × 1	8	double
B	[1 2 3 4]	1 × 4	32	double
C	<3 × 4 double>	3 × 4	96	double
ans	25	1 × 1	8	double

Although we have only introduced variables that are matrices, other types of variables, such as symbolic variables, are possible.

If you prefer to have a less cluttered desktop, you may close any of the windows (except the command window) using the action choices in the upper right-hand corner of each window. You can also personalize which windows you prefer to keep open by selecting Layout in the Home tab bar. If you suppress the workspace window, you can still find out what variables have been defined by using the command

```
whos
```

which returns

Name	Size	Bytes	Class
`A`	1 × 1	8	double
`B`	1 × 4	32	double
`C`	3 × 4	96	double
`ans`	1 × 1	8	double

2.1.4 Current Folder Window

When MATLAB either accesses files or saves information onto your computer, it uses the current folder, which is shown in the **current folder window**. The current folder is shown above the Current Folder window and can be changed to another folder by changing the folder on that line.

2.1.5 Document Window

Double clicking on any variable listed in the workspace window automatically launches a **document window** containing the **array editor**. Values stored in the variable are displayed in a spreadsheet format. You can change values in the array editor, or you can add new values. For example, if you have not already entered the two-dimensional matrix **C**, enter the following command in the command window:

```
C = [1,2,3,4; 10,20,30,40; 5,10,15,20];
```

Placing a semicolon at the end of the command suppresses the output so that it is not repeated back in the command window; however, C should now be listed in the workspace window. Double click it. A document window will open above the workspace window, as shown in Figure 2.2. (Only a portion of the MATLAB screen is shown in Figure 2.2.) You can now add additional values to the C matrix or change existing values.

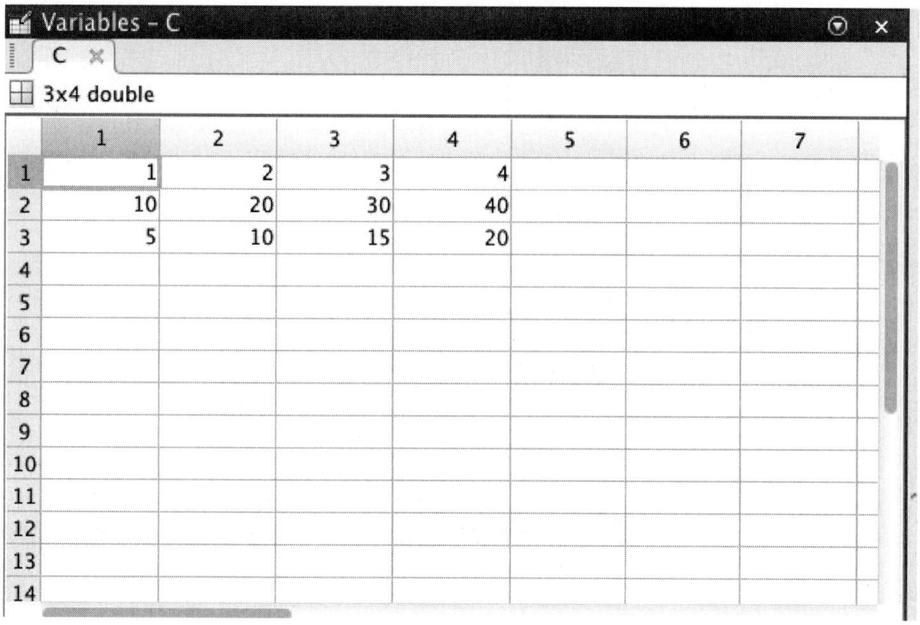

Figure 2.2
Document window with the array editor.

2.1.6 Graphics Window

The **graphics window** launches automatically when you request a graph. To create a simple graph first, create an array of **x** values:

```
x = [1,2,3,4,5];
```

(Remember, the semicolon suppresses the output from this command; however, a new variable **x** appears in the workspace window.) Now create a list of **y** values:

```
y = [1,8,27,64,125];
```

To create a graph, use the **plot** command:

```
plot(x,y)
```

The graphics window opens automatically. (See Figure 2.3.) Notice that a new window label at the top of the Graphics Window identifies this plot as Figure 1. Any additional graphs you create will overwrite Figure 1 unless you specifically command MATLAB to open a new graphics window. MATLAB makes it easy to modify graphs by adding titles, *x* and *y* labels, multiple lines, and more; these options will be discussed more in Chapter 4.

2.1.7 Edit Window

The **edit window** is opened by choosing New Script from the Home tab bar. This window allows you to type and save a series of commands without executing them. You may also open the edit window by typing **edit** at the command prompt.

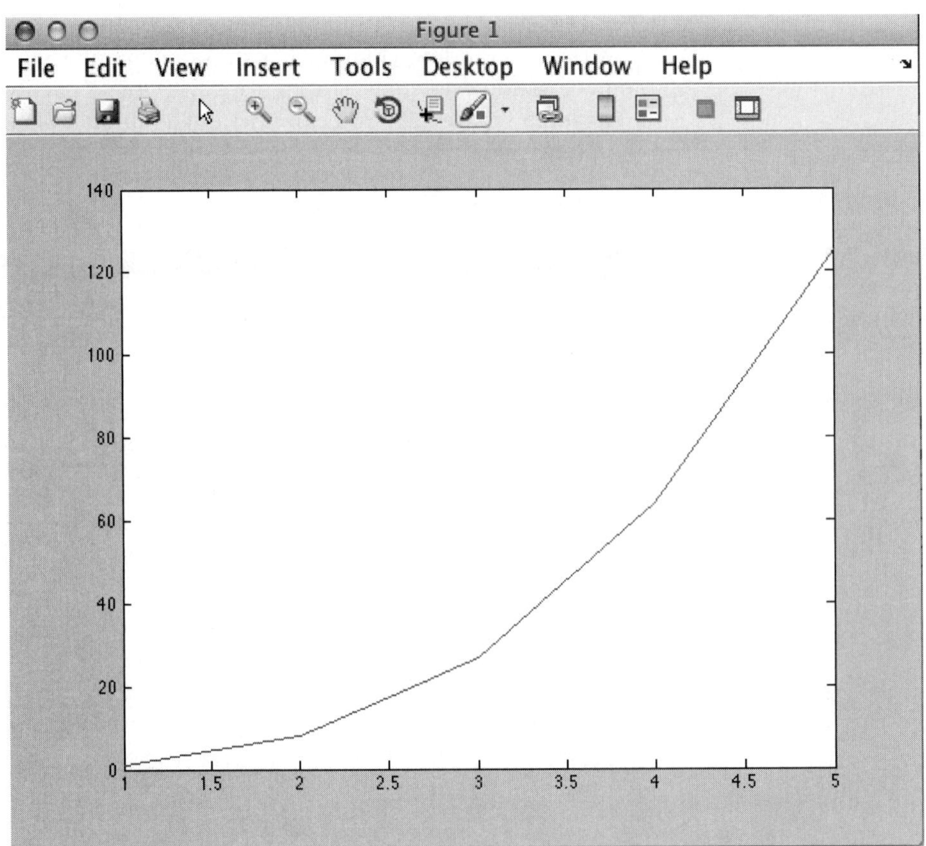

Figure 2.3
MATLAB graphs.

The last section in this chapter will give an example of creating and executing an M-file.

2.2 SIMPLE OPERATIONS

The command window environment is a powerful tool for solving engineering problems. To use it effectively, you will need to understand more about how MATLAB stores information, and then performs operations on that information. In Chapter 3, we cover the wide range of functions that are also available in MATLAB.

2.2.1 Defining Variables

When solving engineering problems, it is important to visualize the data related to the problem. Sometimes the data are just a single number, such as the radius of a circle. Other times, the data may be a coordinate on a plane that can be represented as a pair of numbers, with one number representing the *x*-coordinate and the other number representing the *y*-coordinate. In another problem, we might have a set of four *x-y-z* coordinates that represent the four vertices of a pyramid with a triangular base in a three-dimensional space. We can represent all of these examples using a special type of data structure called a matrix. A **matrix** is a set of numbers arranged in a rectangular grid of rows and columns. Thus, a single point can be considered a matrix with one row and one column—often referred to as a **scalar**. An *x-y* coordinate can be considered a matrix with one row and two columns, and is often called a **vector**. A set of four *x-y-z* coordinates can be considered a matrix with four rows and three columns. Examples are the following:

$$\mathbf{A} = [3.5] \qquad \mathbf{B} = [1.5 \quad 3.1] \qquad \mathbf{C} = \begin{bmatrix} -1 & 0 & 0 \\ 1 & 1 & 0 \\ 1 & -1 & 0 \\ 0 & 0 & 2 \end{bmatrix}$$

Note that the data within a matrix are written inside brackets.

In MATLAB, we assign names to the scalars, vectors, and matrices we use. The following rules apply to these **variable names**:

- Variable names must start with a letter.
- Variable names are case sensitive. The names **time**, **Time**, and **TIME** all represent different variables.
- Other than the first letter, variable names can contain letters, digits, and the underscore (_) character. To test whether a name is a legitimate variable name, use the **isvarname** command. The answer `1` means true, and the answer `0` means false. For example,

```
isvarname Vector
ans =
     1
```

  means that **`Vector`** is a legitimate variable name.
- Variable names can be any length, but only the first *N* characters are used by MATLAB. The value of *N* varies, depending on the version of MATLAB that you are using. For Version 8.2, R2013b, the value of *N* is 63. You can see the value of *N* on your system by typing

```
namelengthmax
```

- Variables cannot have the same name as any of MATLAB's keywords. To see a list of all MATLAB keywords, type

```
iskeyword
```

- MATLAB allows you to use the names of its built-in functions as variable names. This is a dangerous practice since you can overwrite the meaning of a function, such as cos. To check whether a name is a built-in function, use the **`which`** command. For example, typing

```
which cos
```

returns

```
built-in
```

together with the function's directory location.

MATLAB also contains a number of predefined constants and special values that are available to our programs. These values are described in the following list:

`ans`	Represents a value computed by an expression but not stored in a variable name.
`clock`	Represents the current time in a six-element row vector containing year, month, day, hour, minute, and seconds.
`date`	Represents the current date in a character-string format, such as 12-Sep-2013.
`eps`	Represents the floating-point precision for the computer being used. This epsilon precision is the smallest amount with which two values can differ in the computer. For Version 8.2, this value is 2.2204e-16.
`i, j`	Represents the value $\sqrt{-1}$.
`Inf`	Represents infinity, which typically occurs as a result of a division by zero. A warning message will be printed when this value is computed.
`NaN`	Represents Not-a-Number and typically occurs when an expression is undefined, as in the division of zero by zero.
`pi`	Represents π.

2.2.2 Scalar Operations

The arithmetic **operations** between two scalars are shown in Table 2.1. They include addition, subtraction, multiplication, division, and exponentiation. The command

```
a = 1 + 2;
```

Table 2.1 Arithmetic Operations between Two Scalars

Operation	Algebraic Form	MATLAB Form
Addition	$a + b$	**a + b**
Subtraction	$a - b$	**a - b**
Multiplication	$a \times b$	**a*b**
Division	$\frac{a}{b}$	**a/b**
Exponentiation	a^b	**a^b**

should be read as **a** is assigned a value of 1 plus 2, which is the addition of two scalar quantities. Assume, for example, that you have defined **a** in the previous statement and that **b** has a value of 5:

```
b = 5;
```

Then

```
x = a + b
```

will return the following result:

```
x =
    8
```

The equal sign in MATLAB is called the **assignment operator**. The assignment operator causes the result of your calculations to be stored in a computer memory location. In the example above, **x** is assigned a value of eight, and is stored in computer memory. If you enter the variable name

```
x
```

into MATLAB, you get the result

```
x =
    8
```

which should be read as "**x** is assigned a value of 8." If we interpret assignment statements in this way, we are not disturbed by the valid MATLAB statement

```
x = x + 1
```

which, since the value stored in **x** was originally 8, returns

```
x =
    9
```

indicating that the value stored in the memory location named **x** has been changed to 9. Clearly, this statement is not a valid algebraic statement, but is understandable when viewed as an assignment rather than as a statement of equality. The assignment statement is similar to the familiar process of saving a file. When you first save a word processing document, you assign it a name. Subsequently, when you've made changes, you resave your file, but still assign it the same name. The first and second versions are not equal; you've just assigned a new version of your document to an existing memory location.

Because several operations can be combined in a single arithmetic expression, it is important to know the **precedence** of arithmetic operations, or the order in which operations are performed. Table 2.2 contains the precedence of arithmetic operations performed in MATLAB. Note that this precedence follows the standard algebraic precedence rules.

Table 2.2 Precedence of Arithmetic Operations

Precedence	Operation
1	Parentheses, innermost first
2	Exponentiation, left to right
3	Multiplication and division, left to right
4	Addition and subtraction, left to right

To illustrate the precedence of arithmetic operations, assume that we want to calculate the area of a trapezoid, where the base is horizontal and the two edges are vertical. Assume that values have been entered into variables using these commands:

```
base = 5;
height_1 = 12;
height_2 = 6;
```

Now you can compute the area by entering the equation for area in MATLAB:

```
area = 0.5*base*(height_1 + height_2)
```

This equation returns

```
area =
       45
```

Understanding the order of operation is important. Because of the parentheses, MATLAB will first add the two height values together, and then perform the multiplication operations, starting from the left.

Neglecting the parentheses will result in the wrong answer. For example,

```
area = 0.5*base*height_1 + height_2
```

gives

```
area =
      36
```

In this case, MATLAB will first perform the multiplications and then add the result to **height_2**. Clearly, it is important to be very careful when converting equations into MATLAB statements. Adding extras parentheses is an easy way to ensure that computations are performed in the order you want.

If an expression is long, break it into multiple statements. For example, consider the equation

$$f = \frac{x^3 - 2x^2 + x - 6.3}{x^2 + 0.05005x - 3.14}$$

The value could be computed with the following MATLAB statements:

```
numerator = x^3 - 2*x^2 + x - 6.3;
denominator = x^2 + 0.05005*x - 3.14;
f = numerator/denominator;
```

It is better to use several statements that are easy to understand than to use one statement that requires careful thought to figure out the order of operations.

HINT

MATLAB does not read "white space," so it does not matter if you add spaces to your commands. It is easier to read a long expression if you add a space before and after plus and minus signs, but not after multiplication and division signs.

The variables stored in a computer can assume a wide range of values. For most computers, the range extends from 10^{-308} to 10^{308} which should be enough to accommodate most computations. However, it is possible for the result of an expression to be outside of this range. For example, suppose that we execute the following commands:

```
x = 2.5e200;
y = 1.0e200;
z = x*y
```

MATLAB responds with

```
z =
   Inf
```

because the answer (2.5e400) is outside of the allowable range. This error is called **exponent overflow**, because the exponent of the result of an arithmetic operation is too large to store in the computer's memory.

Exponent underflow is a similar error, caused by the exponent of the result of an arithmetic operation being too small to store in the computer's memory. Using the same allowable range, we obtain an exponent underflow with the commands

```
x = 2.5e-200;
y = 1.0e200;
z = x/y
```

which together return

```
z =
   0
```

The result of an exponent underflow is zero.

We also know that the division by zero is an invalid operation. If an expression results in a division by zero, the result of the division is infinity:

```
z = y/0
z =
    Inf
```

MATLAB may also print a warning telling you that division by zero is not possible.

EXAMPLE 2.1

WIND TUNNEL ANALYSIS

Wind tunnels are used to evaluate high performance aircraft. (See Figure 2.4.) To interpret wind tunnel data, the engineer needs to understand how gases behave. The basic equation describing gas properties is the ideal gas law,

$$PV = nRT$$

where

P = pressure, kPa
V = volume, m^3
n = number of kmoles of gas in the sample
R = ideal gas constant, 8.314 kPa m^3/kmole K
T = temperature, expressed on an absolute scale (i.e., in degrees K)

(*continued*)

Figure 2.4 Wind tunnels used to test aircraft designs.

In addition, we know that the number of kmoles of gas is equal to the mass of gas divided by the molar mass (also known as the molecular weight); that is,

$$n = m/\text{MW}$$

where

m = mass, kg
MW = molar mass, kg/kmole

Different units can be used in the equations, if the value of R is changed accordingly. Assume that the volume of air in the wind tunnel is 1000 m^3. Before the wind tunnel is turned on, the temperature of the air is 300 K and the pressure is 100 kPa. The molar mass (molecular weight) of air is approximately 29 kg/kmole. Find the mass of air in the wind tunnel.

SOLUTION

1. Problem Statement

Find the mass of air in the wind tunnel.

2. Input/Output Description

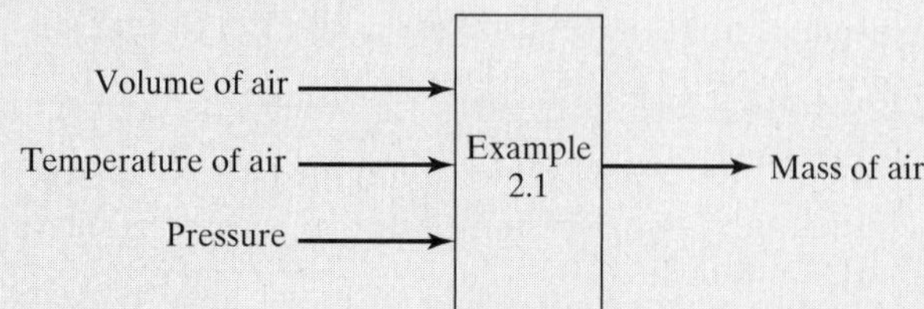

3. Hand Example

Working the problem by hand (or with a calculator) allows you to outline an algorithm, which you can translate to MATLAB code later. You should choose simple data that make it easy to check your work. In this example, we want to solve the ideal gas law for n, and plug in the given values. This results in

$$\begin{aligned} n &= PV/RT \\ &= (100\ \text{kPa} \times 1000\ \text{m}^3)/(8.314\ \text{kPa m}^3/\text{kmole K} \times 300\ \text{K}) \\ &= 40.0930\ \text{kmoles} \end{aligned}$$

Convert moles to mass by multiplying by the molar mass:

$$m = n \times \text{MW} = 40.0930\ \text{kmoles} \times 29\ \text{kg/kmole}$$

$$m = 1162.70\ \text{kg}$$

4. MATLAB Solution

The solution to this problem is really just entering data, and then computing two equations. The command window in Figure 2.5 shows the commands, together with their corresponding MATLAB output. Notice also the use of parentheses in the denominator computation; they are necessary for the correct calculation of the denominator value.

5. Testing

In this case, comparing the result to the hand result is sufficient. More complicated problems solved in MATLAB should use a variety of input data to confirm that your solution works in a variety of cases. Notice that the variables defined in the

Figure 2.5
Wind tunnel air mass.

```
Command Window
>> P = 100;
>> V = 1000;
>> R = 8.314;
>> T = 300;
>> MW = 29;
>> n = P*V/(R*T)

n =

   40.0930

>> m = n*MW

m =

  1.1627e+003

fx >>
```

(continued)

command window are listed in the workspace window. Also notice that the command history lists the commands executed in the command window. If you were to scroll up in the command history window, you would see commands from previous MATLAB sessions. All of these commands are available for you to move to the command window.

2.2.3 Element-By-Element Operations

Using MATLAB simply as a calculator can be useful, but its real strength is in matrix manipulations. (Remember that when we use the term matrix, we can be referring to a scalar, a vector, or a matrix with rows and columns.) As described previously, the easiest way to define a matrix is to use a list of numbers called an explicit list. The command

```
X = [1,2,3,4];
```

defines a row vector with four values. Recall that when defining this vector, you may either list the values with or without commas. A new row is indicated by a semicolon, so that a column vector is specified as

```
Y = [1; 2; 3; 4];
```

and a matrix that contains both rows and columns would be created with the statement

```
A = [1,2,3,4; 2,3,4,5; 3,4,5,6];
```

HINT

You can also keep track of how many values you have entered into a matrix if you enter each row on a separate line. Thus, another way to enter values into the matrix **A** above would be:

```
A = [1,2,3,4;
     2,3,4,5;
     3,4,5,6]
```

While a complicated matrix might have to be entered by hand, evenly spaced matrices can be entered much more readily. The command

```
B = 1:5;
```

or the command

```
B = [1:5];
```

defines a row matrix with the five values 1, 2, 3, 4, 5. The square brackets are optional. The default increment is 1, but if you want to use a different increment, put it between the first and final values. For example,

```
C = 1:2:5
```

indicates that the increment between values will be 2 and displays

```
C =
    1  3  5
```

If you want MATLAB to calculate the increment between values, you can use the **linspace** command. Specify the initial value, the final value, and how many total values you want. For example,

```
D = linspace(1,10,3)
```

initializes a vector with three values, evenly spaced between 1 and 10, as shown below:

```
D =
    1.0000   5.5000   10.0000
```

Matrices can be used in many calculations with scalars. If **A** = [1 2 3], we can add 5 to each value in the matrix with the calculation

```
B = A + 5
```

which displays

```
B =
    6  7  8
```

This works well for addition and subtraction; however, multiplication and division are slightly different. In matrix mathematics, the multiplication operator has a very specific meaning. If you want to do an element-by-element multiplication, the operator must be preceded by a period. For example,

```
A.*B
```

results in the first element of **A** multiplied by the first element of **B**, the second element of **A** multiplied by the second element of **B**, and so on. For the example of **A** (which is **[1,2,3]**) and **B** (which is **[6,7,8]**),

```
A.*B
```

returns

```
ans =
      6  14  24
```

(Be sure to do the math to convince yourself why these are the correct answers.)

Using just an asterisk (instead of a period and an asterisk) specifies matrix multiplication which is discussed in Chapter 6. If you get error messages in computation statements, be sure to see if you have forgotten to include the period for **element-by-element operations.**

The same syntax holds for element-by-element division and exponentiation, as shown in these statements:

```
A./B
A.^2
```

As an exercise, predict the values resulting from the preceding two expressions, then test out your predictions by executing the commands in MATLAB.

The matrix capability of MATLAB makes it easy to do repetitive calculations. For example, assume you have a list of angles in degrees that you would like to convert to radians. First, put the values into a matrix. For angles of 10, 15, 70, and 90, enter

```
D = [10,15,70,90];
```

To change the values to radians, you must multiply by $\pi/180$:

```
R = D*pi/180;
```

This command returns a matrix **R**, with the values in radians. (In the next chapter, we will show you easier ways to convert degrees to radians, or from radians to degrees, using built-in functions.)

HINT

The value of π is built into MATLAB as a floating point number, called **pi**. Because π is an irrational number, it cannot be expressed exactly with a floating point representation, and the MATLAB constant, **pi**, is really an approximation.

Another useful matrix operator is transposition. The **transpose operator** basically changes rows to columns or vice versa. For example, using the matrix **D** defined above,

```
D'
```

displays

```
ans =
      10
      15
      70
      90
```

This makes it easy to create tables. For example, to create a table of degrees to radians, enter

```
table = [D',R']
```

which tells MATLAB to create a matrix named **table**, where the first column is **D'**, and the second column is **R'**:

```
table =
        10.0000   0.1745
        15.0000   0.2618
        70.0000   1.2217
        90.0000   1.5708
```

EXAMPLE 2.2

SCIENTIFIC DATA CONVERSION

Scientific data, such as that collected from wind tunnels, are usually in SI (system international) units. However, much of the manufacturing infrastructure in the United States has been tooled in English (sometimes called American Engineering or American Standard) units. Engineers need to be fluent in both systems, and especially careful when sharing data with other engineers. Perhaps the most notorious example of unit confusion problems occurred in the flight of the *Mars Climate Orbiter*, the second flight of the NASA Mars Surveyor Program. (See Figure 2.6.) The spacecraft burned up in the orbit of Mars in September 1999 because of a look-up table embedded in the spacecraft's software. The table, probably generated from wind tunnel testing, used pounds force (lbf), when the program expected values in newtons (N).

Figure 2.6
Mars Climate Orbiter.

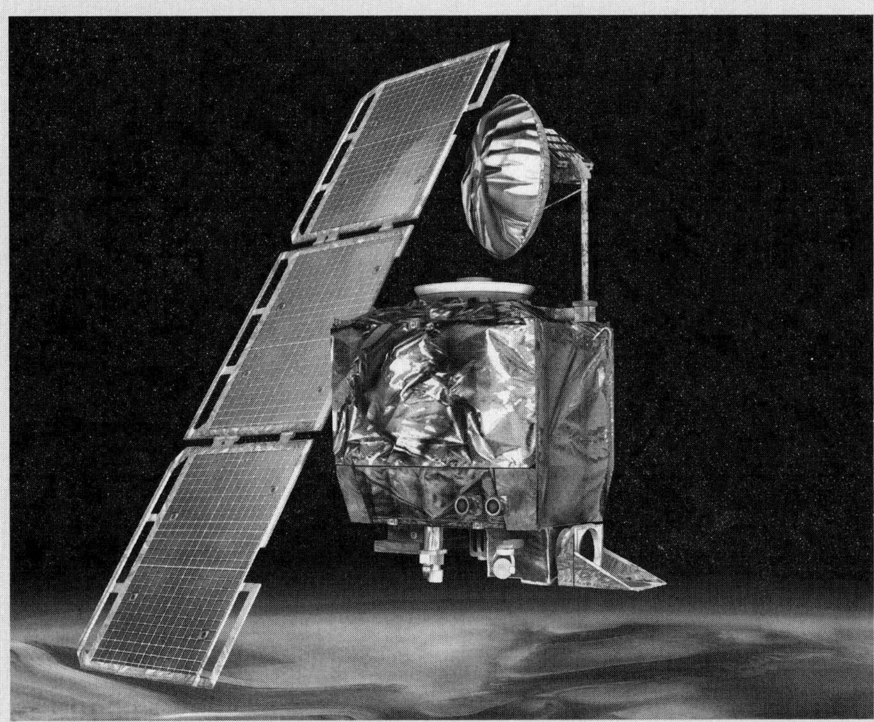

Use MATLAB to create a conversion table of pounds force (lbf) to newtons (N). Your table should start at 0 and go to 1000 lbf, at 100 lbf intervals. Use the conversion

$$1 \text{ lbf} = 4.4482216 \text{ N}$$

1. Problem Statement

Create a table converting pound force (lbf) to newtons (N).

2. Input/Output Description

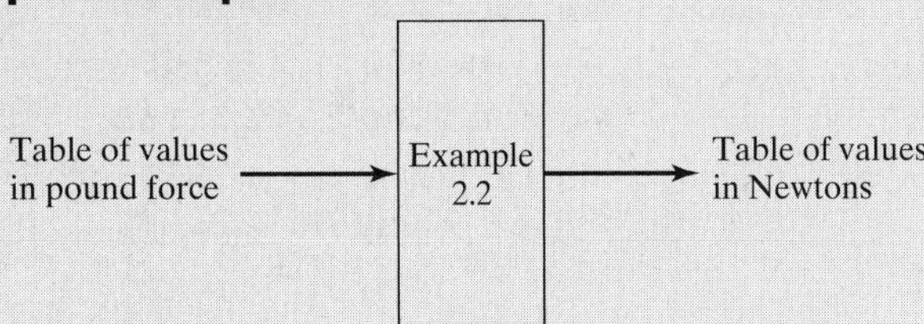

(continued)

3. Hand Example

Here are a few conversions that we compute by hand so that we can then compare them to the entries in the table generated by MATLAB:

$$0 \times 4.4482216 = 0$$

$$100 \times 4.4482216 = 444.82216$$

$$1000 \times 4.4482216 = 4448.2216$$

4. MATLAB Solution

It is always a good idea to clear the command space and the workspace before beginning a new problem solution. The **clc** command clears the command window and the command **clear** removes all variables from memory. The command **clf** clears the current figure and thus clears the graph window.

See Figure 2.7 for the commands entered, and the table computed by MATLAB. Notice in the workspace window that lbf and N are 1×11 matrices, and that **ans** (which is where the table we created is stored) is an 11×2 matrix. The output from the first two commands was suppressed by adding a semicolon at the end of each line. It would be very easy to create a table with more entries by changing the increment to 10 or even to 1. Also notice that you will need to multiply the table results by 1000 to get the correct answers. MATLAB tells you this is necessary directly above the table.

Figure 2.7
Command window with solution to Example 2.1.

Command Window

```
>> lbf = 0:100:1000;
>> N = 4.4482216*lbf;
>> [lbf',N']

ans =

  1.0e+003 *

         0         0
    0.1000    0.4448
    0.2000    0.8896
    0.3000    1.3345
    0.4000    1.7793
    0.5000    2.2241
    0.6000    2.6689
    0.7000    3.1138
    0.8000    3.5586
    0.9000    4.0034
    1.0000    4.4482
```

5. Testing

Comparing the results of the MATLAB solution to the hand solution shows the same results. Once we have verified that our solution works, it is easy to use the same algorithm to create other conversion tables. For example, modify this example to create a conversion table of newton (N) to pound force, with an increment of 10 N, from 0 N to 1000 N.

2.2.4 Additional Ways to Define Matrices

As you solve more and more complicated problems with MATLAB, you will find that you need to combine small matrices into larger matrices, extract information from large matrices, create very large matrices, and use matrices with special properties. In this section, we demonstrate some of the techniques for performing these operations.

We know that a matrix can be defined by typing in a list of numbers enclosed in square brackets. The numbers can be separated by spaces or commas; we prefer to use commas to make the numbers easier to read. New rows are indicated with a semicolon within the brackets:

```
A = [3.5];
B = [1.5, 3.1];
C = [-1,0,0; 1,1,0; 0,0,2];
```

A matrix can also be defined by listing each row on a separate line, as in the following set of MATLAB commands:

```
C = [-1,0,0;
     1,1,0;
     1,-1,0;
     0,0,2];
```

If there are too many numbers in a row to fit in one line, you can continue the statement in the next line, but a comma and an **ellipsis** (three dots) are needed at the end of the line to indicate that the row is to be continued. (You can also use the ellipsis to continue long assignment statements in MATLAB.) For example, if we want to define **F** with 10 values, we could use either of the following statements:

```
F = [1,52,64,197,42,-42,55,82,22,109];
```

or

```
F = [1,52,64,197,42,-42, ...
    55,82,22,109];
```

MATLAB also allows you to define a matrix by using another matrix that has already been defined. For example, the following statements

```
B = [1.5, 3.1];
S = [3.0, B]
```

define a new vector **S**, and its value is displayed as the following:

```
S =
    3.0000   1.5000   3.1000
```

Similarly,

```
T = [1,2,3; S]
```

displays

```
T =
    1.0000  2.0000  3.0000
    3.0000  1.5000  3.1000
```

We can also change values in a matrix, or include additional values, by using a reference to specific locations. Thus, the command

```
S(2) = -1.0;
```

changes the second value in the matrix `S` from 1.5 to −1. If you then type the matrix name `S` into the command window—that is,

```
S
```

then MATLAB displays

```
S =
    3.0000  -1.0000  3.1000
```

We can also extend a matrix by defining new elements. If we execute the command

```
S(4) = 5.5;
```

we extend the matrix `S` to four elements instead of three. If we define element `S(8)` using the following statement:

```
S(8) = 9.5;
```

then the matrix `S` will have eight values, and the values of `S(5)`, `S(6)`, and `S(7)` will be set to zero. Thus, entering the matrix name in the command line will display the following:

```
S =
    3.0000  -1.0000  3.1000  5.5000  0  0  0  9.5000
```

The **colon operator** is a very powerful operator for defining new matrices and modifying existing matrices. An evenly spaced matrix can be defined with the colon operator. Thus,

```
H = 1:8
```

will define a matrix and display it as follows:

```
H =
    1  2  3  4  5  6  7  8
```

The default spacing for the colon operator is 1. However, when colons are used to separate three numbers, the middle value becomes the spacing. For example,

```
time = 0.0:0.5:2.0
```

will define the vector and display it as follows:

```
time =
       0  0.5000  1.0000  1.5000  2.0000
```

The colon operator can also be used to extract data from matrices, which becomes very useful in data analysis. When a colon is used in a matrix reference in place of a specific subscript, the colon represents the entire row or column. To illustrate, assume that we define a matrix with the following statement:

```
M = [1,2,3,4,5;
     2,3,4,5,6;
     3,4,5,6,7];
```

We can then extract the first column with this command

```
x = M(:,1)
```

which defines a matrix **x**, and displays the following:

```
x =
    1
    2
    3
```

You can extract any of the columns in a similar manner. To extract the fourth column, we can use the following statement:

```
y = M(:,4)
```

The output displayed is the following:

```
y =
    4
    5
    6
```

Similarly, to extract a row, use the following:

```
z = M(1,:)
```

which displays

```
z =
    1  2  3  4  5
```

In all of the preceding examples, read the colon as "all of the rows," or "all of the columns." You do not have to extract an entire row or an entire column. The colon operator can also be used to select values "from row i to row j" or "from column m to column n." For example, to extract the two bottom rows of the **M** matrix, we can enter the following:

```
w = M(2:3,:)
```

which defines a new matrix **w** and displays its contents:

```
w =
    2  3  4  5  6
    3  4  5  6  7
```

Similarly, to extract just the four numbers in the lower right-hand corner of matrix **M**, use

```
w = M(2:3,4:5)
```

which displays

```
w =
     5   6
     6   7
```

In MATLAB, it is valid to have a matrix that is empty. For example, the following statements will each generate an **empty matrix**:

```
a = [];
b = 4:-1:5;
```

Finally, using the matrix name with a single colon, as in

```
x = M(:);
```

transforms **M** into one long column matrix that is formed from selecting column 1, followed by column 2, then column 3, then column 4, and then column 5. It is important to recognize that any matrix can be represented as a single column matrix using this notation. Therefore, if you reference a matrix using only a single subscript, as in **M(8)**, this reference will always assume that you are using the single column matrix representation. Thus, in this example, **M(8)** is the value in the second row and third column, or a 4.

2.3 OUTPUT OPTIONS

MATLAB gives us a number of options for displaying information. We can choose to have numbers displayed in a variety of forms (for example, use a specific number of decimal positions or use a scientific format). We can also use MATLAB statements within our programs to display information in different ways. In this section we first present the ways in which numbers can be displayed, and then we present two MATLAB commands that give us additional control over the output of information.

2.3.1 Number Display

In this section, we present the options that we have to display information in MATLAB. These options include a scientific notation, and then a number of ways of displaying numeric information.

Although you can enter any number in decimal notation, it is not always the best way to represent very large or very small numbers. For example, a number that is used frequently in chemistry is Avogadro's constant, whose value to four significant digits is 602,200,000,000,000,000,000,000. The diameter of an iron atom is approximately 140 picometers, which is .000000000140 meters. **Scientific notation** expresses a value as a number between 1 and 10 (the mantissa) multiplied by a power of 10 (the exponent). In scientific notation, Avogadro's number becomes 6.022×10^{23}, and the diameter of an iron atom becomes 1.4×10^{-10} meters. In MATLAB, values in scientific notation are designated with an *e* between the mantissa and the exponent, and this notation is referred to as an **exponential notation**. For example,

```
Avogadros_constant = 6.022e23;
Iron_diameter = 1.4e-10;
```

It is important to omit blanks between the mantissa and the exponent. For example, MATLAB will interpret

```
6.022      e23
```

as two different values (6.022 and 10^{23}).

When elements of a matrix are displayed in MATLAB, integers are always printed as integers. However, values with decimal fractions are printed using a default format that shows four decimal digits. Thus,

```
A = 5
```

returns

```
A =
    5
```

but

```
A = 51.1
```

returns

```
A =
    51.1000
```

MATLAB allows you to specify other formats that show more significant digits. For example, to specify that we want values to be displayed in a decimal format with 14 decimal digits, we use the command

```
format long
```

which changes all subsequent displays. For example,

```
A
```

now returns

```
A =
    51.10000000000000
```

We can return the format to four decimal digits by using the command

```
format short
A
A =
    51.1000
```

Two decimal digits are displayed when the format is specified as **format bank**. No matter what display format you choose, MATLAB uses double precision floating-point numbers in its calculations. Exactly how many digits are used in these calculations depends upon your computer. However, changing the display format does not change the accuracy of your results. When numbers become too large or too small for MATLAB to display using the default format, the program automatically expresses them in scientific notation. For example, if you enter Avogadro's constant into MATLAB in decimal notation,

```
avo = 602200000000000000000000
```

the program returns

```
avo =
      6.0220e+023
```

You can force MATLAB to display all numbers in scientific notation with **format short e** (with 5 significant digits) or **format long e** (with 14 significant digits). Thus,

```
format short e
x = 10.356789
```

returns

```
x =
    1.0357e+01
```

Another format command is **format +**. When a matrix is displayed with this format, the only characters printed are plus and minus signs. If a value is positive, a plus sign will be displayed; if a value is negative, a minus sign will be displayed. If a value is zero, nothing will be displayed. This format allows us to view a large matrix in terms of its signs:

```
format +
B = [1,-5,0,12; 10005,24,-10,4]
B =
    +- +
    ++-+
```

For long and short formats, a common scale factor is applied to the entire matrix if the elements become very large or very small. This scale factor is printed along with the scaled values.

Finally, the command **format compact** suppresses many of the line feeds that appear between matrix displays and allows more lines of information to be seen together on the screen. The command **format loose** will return the command window to the less compact display mode. The examples in this text use the compact format to save space. To return to the default format, use the command **format** without any additional information. Table 2.3 contains a summary of the numeric display formats.

2.3.2 Display Function

The display (**disp**) function can be used to display the contents of a variable without printing the variable's name. Consider these commands:

```
x = 1:5;
disp(x)
```

Table 2.3 Numeric Display Formats

MATLAB Command	Display	Example
format short	4 decimal digits	15.2345
format long	14 decimal digits	15.23453333333333
format short e	4 decimal digits	1.5234e+01
format long e	15 decimal digits	1.523453333333333e+01
format bank	2 decimal digits	15.23
format +	+, −, blank	+

returns

```
1  2  3  4  5
```

The display command can also be used to display a string (text enclosed in single quote marks). This can be useful when displaying information from M-files, which are discussed later in this chapter. An example of using this command to give a header to a set of data that is printed from an M-file is shown below:

```
disp('The values in the x matrix are:');
disp(x);
```

These statements return the following:

```
The values in the x matrix are:

1  2  3  4  5
```

The semicolon at the end of the **disp** statement is optional. Notice that the output from the two **disp** statements is displayed on separate lines.

2.3.3 Formatted Output

The **fprintf** function gives you even more control over the output than you have with the **disp** function. In addition to displaying both text and matrix values, you can specify the format to be used in displaying the values, and you can specify when to skip to a new line. If you are a C programmer, you will be familiar with the syntax of this function. With few exceptions, the MATLAB **fprintf** functions use the same formatting specifications as the C **fprintf** function.

The general form of this command contains a string and a list of matrices:

```
fprintf(format-string,variables,...)
```

Consider the following example:

```
temp = 98.6;
fprintf('The temperature is %f degrees F \n',temp);
```

The character string in single quote marks, which is the first argument inside the **fprintf** function, contains the percent character to begin a formatting operator that will specify the format of the value to be printed at that location. In this example, the formatting is defined by the conversion character, **f**, which tells MATLAB to display the temperature value in a default fixed-point format. Note in the output generated below that the default precision in the printed output has six digits after the decimal point:

```
The temperature is 98.600000 degrees F
```

The final character sequence in the formatting string, **\n**, inserts a special character in the printed output that causes MATLAB to start a new line. Format operations usually end with **\n** so that subsequent output will start on a new line, rather than the current one.

There are many other conversion characters (such as **%e** and **%g**) and special characters that can be used in the formatting operator to produce the output that is desired. Some of these characters are shown in Tables 2.4 and 2.5.

You can further control how the variables are displayed by using the optional width field and precision field with the format command. The **width field** controls

Table 2.4 Type Field Format

Type Field	Result
%f	fixed point, or decimal notation
%e	exponential notation
%g	whichever is shorter, %f or %e

Table 2.5 Special Format Commands

Format Command	Resulting Action
\n	linefeed
\r	carriage return (similar to linefeed)
\t	tab
\b	backspace

the minimum number of characters to be printed. It must be a positive decimal integer. The **precision field** is preceded by a period and specifies the number of decimal places after the decimal point for exponential and fixed-point types. For example, **%8.2f** specifies that the minimum total width available to display your result is 8 digits, two of which are after the decimal point:

```
fprintf('The temperature is %8.2f degrees F\n',temp);
```

returns

```
The temperature is  98.60 degrees F
```

Many times when you use the **fprintf** function, your variable will be a matrix. For example,

```
temp = [98.6, 100.1, 99.2];
```

MATLAB will repeat the string in the **fprintf** command until it uses all of the values in the matrix:

```
fprintf('The temperature is %8.2f degrees F\n',temp);
```

returns

```
The temperature is   98.60 degrees F
The temperature is  100.10 degrees F
The temperature is   99.20 degrees F
```

If the variable is a two-dimensional matrix, MATLAB uses the values one column at a time, going down the first column, then the second, and so on. Here is a more complicated example:

```
patient = 1:3;
temp = [98.6, 100.1, 99.2];
```

Combine these two matrices:

```
history = [patient; temp]
```

returns

```
history =
            1.0000      2.0000     3.0000
           98.6000    100.1000    99.2000
```

Now we can use the **fprintf** function to create a table that is easier to interpret:

```
fprintf('Patient %1.0f had a temperature of %7.2f \n',history)
```

sends the following output to the command window:

```
Patient  1 had a temperature of   98.60
Patient  2 had a temperature of  100.10
Patient  3 had a temperature of   99.20
```

As you can see, the **fprintf** function allows you to have a great deal of control over the output form.

2.4 SAVING YOUR WORK

Working in the command window is similar to performing calculations on your scientific calculator. When you turn off the calculator, or when you exit the program, your work is gone. It is possible to save the values of the variables that you defined in the command window and that are listed in the workspace window. Although this may be useful, it is more likely that you will want to save the list of commands that generated your results. We will show you how to save and retrieve variables (the results of the assignments you made and the calculations you performed) to **MAT-files** or to **DAT-files**. Then, we will show you how to generate **M-files**, or **script files**, containing MATLAB commands that are created in the edit window. These files can then be retrieved at a later time to give you access to the commands, or programs, that are contained in them. All example programs in the rest of this text will be generated as M-files so that you have the flexibility of working on the programs, and coming back at a later time to continue your work without reentering the commands.

2.4.1 Saving Variables

To preserve the variables you created in the command window (check the workspace window on the right-hand side of the MATLAB screen for the list of variables) between sessions, you must save the contents of the workspace window to a file. The default format is a binary file called a MAT-file. To save the workspace (remember, this is just the set of variables, not the list of commands in the command window) to a file, use the **save** command:

```
save file_name
```

Although **save** is a MATLAB command, **file_name** is a user-defined file name. The file name can be any name you choose, as long as it conforms to the variable naming conventions for MATLAB. If you execute the **save** command without a filename, MATLAB names the file **matlab.mat**. You can also choose Save Workspace from the Home tab, which will then prompt you to enter a file name for your data.

To restore the values of variables to the workspace, type

```
load file_name
```

Again, **load** is a MATLAB command, but **file_name** is the user-defined file name. If you just type **load**, MATLAB will look for the default **matlab.mat** file.

The file you save will be stored in the current directory. To illustrate, type

```
clear, clc
```

This will clear the workspace and the command window. Verify that the workspace is empty by checking the workspace window, or by typing

```
whos
```

Now define several variables, such as:

```
A = 5;
B = [1,2,3];
C = [1,2; 3,4];
```

Check the workspace window once again to confirm that the variables have been stored. Now, save the workspace to a file called **my_example_file**:

```
save my_example_file
```

Confirm that this new file has been stored in the current directory. If you prefer to save the file to another directory (for instance, onto a thumb drive), use the browse button to navigate to the directory of your choice. Remember that in a public computer lab, the current directory is probably purged after each user logs off the system.

Now, clear the workspace and command window again by typing

```
clear, clc
```

The workspace window should be empty. Now load the file back into the workspace:

```
load my_example_file
```

Again, the file you want to load must be in the current directory, or else MATLAB will not be able to find it. In the command window, type the variable names. You should see the names and the values of the variables displayed.

MATLAB can also store individual matrices or lists of matrices into the current directory via the command

```
save file_name variable_list
```

where **file_name** is the user-defined file name where you wish to store the information, and **variable_list** is the list of variables to be stored in the file. For example,

```
save file1 A B
```

would save just the variables **A** and **B** into **file1.mat**.

If your saved data will be used by a program other than MATLAB (such as C or C++), the MAT format is not appropriate, because MAT files are unique to MATLAB. The **ASCII** format is standard between computer platforms and is more appropriate if you need to share files. MATLAB allows you to save files as ASCII files by modifying the save command:

```
save file_name.dat variable_list -ascii
```

The command **-ascii** tells MATLAB to store the data in a standard text format. ASCII files should be saved into a DAT file instead of a MAT file, so be sure to add **.dat** to your file name—if you do not, it will default to **.mat**.

For example, to create the matrix **Z** and save it to the file **data_2.dat** in text format, use the following commands:

```
Z = [5,3,5; 6,2,3];
save data_2.dat Z -ascii
```

This command causes each row of the matrix **Z** to be written to a separate line in the data file. You can view the **data_2.dat** file by double clicking the file name in the current directory window.

To retrieve data from an ASCII DAT file, enter the **load** command followed by the file name. This will cause the information to be read into a matrix with the same name as the data file.

2.4.2 Creating and Using M-Files

In addition to providing an interactive computational environment using the command window, MATLAB can also be used as a powerful programming language. The MATLAB commands that you write represent a program that you can create and save in a file called an **M-file**. An M-file is a text file similar to a C or C++ source-code file. An M-file can be created and edited using MATLAB or you can use another text editor of your choice. The MATLAB editing window is shown in Figure 2.8. To use the MATLAB editing window, choose New Script from the Home tab bar. If you choose **a different text editor, make sure the files you save are in the ASCII format**.

When you save an M-file, it is stored in the current directory. You will need to name your file with a MATLAB variable name; that is, the name must start with a letter and contains only letters, numbers, and the underscore (_). Spaces are not allowed.

There are two types of M-files, called scripts and functions; we discuss scripts here and Chapter 3 will discuss functions. A **script** M-file is simply a list of MATLAB statements that are saved in a file (with a **.m** file extension). The script has access to workspace variables. Any variables created in the script are accessible to the workspace when the script finishes. A script created in the MATLAB editor window can be executed by saving the file and selecting the run option from the Editor tab of the Editor window. (See Figure 2.8.) Alternately, a script can be executed by typing a filename or by using the **run** command from the command window.

For example, assume you have created a script file named **example1.m**. You can either run the script from the edit window or use one of the three equivalent ways of executing the script from the command window, as shown in Table 2.6.

You can find out what M-files are in the current directory by typing

```
what
```

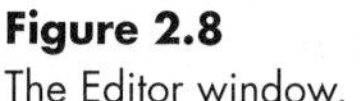

Figure 2.8
The Editor window.

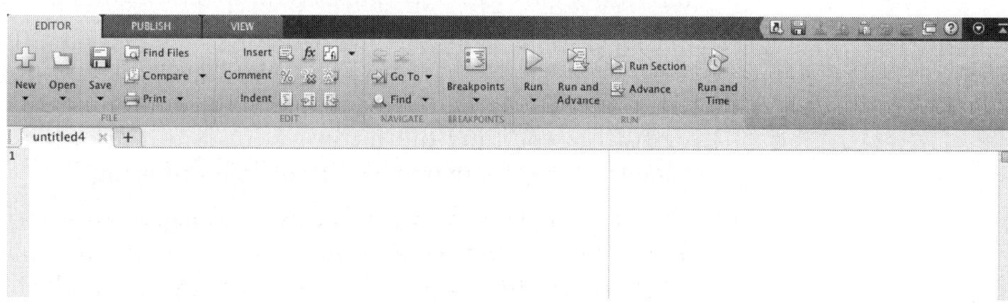

Table 2.6 Executing M-Files from the Command Window

MATLAB Command	Comments
`example1`	Type the file name. The `.m` file extension is assumed.
`run example1`	Use the `run` command with the file name.
`run('example1')`	This method uses the functional form of the `run` command.

into the command window. You can also simply browse through the current directory by looking in the current directory window.

Using M-files allows you to work on a project and to save the list of commands for future use. Because you will be using these files in the future, it is a good idea to add comments to the commands to help you remember the step in the solution, or algorithm. **Comments** are indicated by a percent sign, and they can be on a line by themselves, or at the end of a command line. MATLAB will ignore any comments when the program is executed. We now present an example in which the solution is developed as an M-file. Most of the problem solutions in the rest of this text will be developed as M-files. We suggest that you use M-files when you are using MATLAB for homework solutions (in this and other courses) because you often will want to go back and rerun the program with new data, or you will want to make small changes in the statements. If you have not used an M-file to store the program, you will need to reenter the complete program when you need it in another work session.

EXAMPLE 2.3

UNDUCTED FAN ENGINE PERFORMANCE

An advanced turboprop engine, called an **unducted fan** (UDF), is one of the promising new propulsion technologies being developed for future transport aircraft. (See Figure 2.9.) Turboprop engines, which have been in use for decades, combine the power and reliability of jet engines with the efficiency of propellers. They are a significant improvement over earlier piston-powered propeller engines. Their application has been limited to smaller commuter-type aircraft, however, because they are not as fast or powerful as the fan-jet engines used on larger airliners. The UDF engine employs significant advancements in propeller technology, narrowing the performance gap between turboprops and fan-jets. New materials, blade shapes, and higher rotation speeds enable UDF-powered aircraft to fly almost as fast as fan-jets, and with greater fuel efficiency. The UDF is also significantly quieter than the conventional turboprop.

During the test flight of a UDF-powered aircraft, the pilot has set the engine power level at 40,000 N, which causes the 20,000 kg aircraft to attain a cruise speed of 180 m/s. The engine throttles are then set to a power level of 60,000 N, and the aircraft begins to accelerate. As the speed of the plane increases, the aerodynamic drag increases in proportion to the square of the air speed. Eventually, the aircraft reaches a new cruise speed, where the thrust from the UDF engines is just offset by the drag. The equations used to estimate the velocity and acceleration of the aircraft from the time the throttle is reset to the time the plane reaches new cruise speed (at approximately 120 s) are the following:

$$\text{velocity} = 0.00001\ \text{time}^3 - 0.00488\ \text{time}^2 + 0.75795\ \text{time} + 181.3566$$

$$\text{acceleration} = 3 - 0.000062\ \text{velocity}^2$$

Figure 2.9
An unducted fan (UDF) engine.

Write a MATLAB program, using a script M-file, that calculates the velocity and acceleration of the aircraft at times from 0 to 120 seconds, and at increments of 10 seconds. Assume that time zero represents the point at which the power level was increased. Display the results in a table of time, velocity, and acceleration.

SOLUTION

1. Problem Statement

Calculate the velocity and acceleration, using a script M-file.

2. Input/Output Description

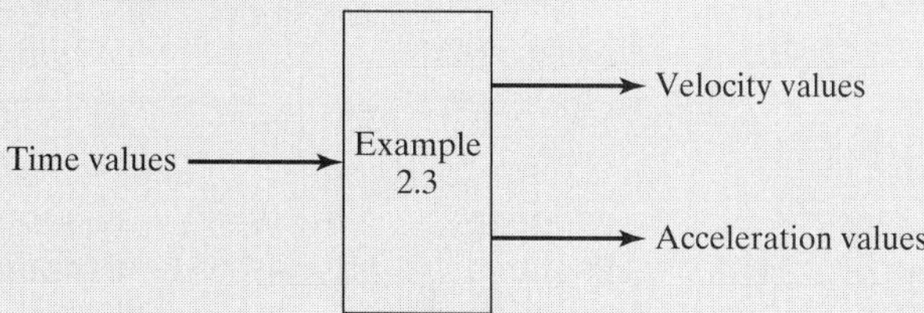

3. Hand Example

Solve the equations stated in the problem for time = 100 seconds;

$$\begin{aligned}\text{velocity} &= 0.00001\ \text{time}^3 - 0.00488\ \text{time}^2 + 0.75795\ \text{time} + 181.3566 \\ &= 218.35\ \text{m/s}\end{aligned}$$

(*continued*)

$$\begin{aligned}\text{acceleration} &= 3 - 0.000062\ \text{velocity}^2\\ &= 0.04404\ \text{m/s}^2\end{aligned}$$

4. MATLAB Solution

Create a new script M-file by choosing New Script from the Home tab bar. Enter these commands:

```
%------------------------------------------------------------
% Example 2_3 This program generates velocity
% and acceleration values for a UDF aircraft test.
%
clear, clc
%
% Define the time matrix.
time = 0:10:120;
%
% Calculate the velocity and acceleration values.
velocity = 0.00001*time.^3 - 0.00488*time.^2 ...
           + 0.75795*time + 181.3566;
acceleration = 3 - 6.2e-05*velocity.^2;
%
% Display the results in a table.
disp('Time, Velocity, Acceleration Values');
disp([time',velocity',acceleration']);
%------------------------------------------------------------
```

Save the file with a name of your choice. Remember that the file name needs to be a legitimate variable name. For example, you might save the file as **Example2_3** or **ch2_ex3**.

5. Testing

Execute the file by selecting the Run icon from the Editor tab. The following results will be displayed in the command window:

```
Time, Velocity, Acceleration Values
        0     181.3566     0.9608
  10.0000     188.4581     0.7980
  20.0000     194.6436     0.6511
     :
 100.0000     218.3516     0.0440
 110.0000     218.9931     0.0266
 120.0000     219.3186     0.0178
```

The command window does not contain the commands executed in the script M-file. The hand example values match those computed by MATLAB. It will now be easy to modify the M-file to make changes to the program to solve other similar problems.

2.4.3 User Input from M-Files

We can create more general programs by allowing the user to input variable values from the keyboard while the program is running. The **input** command allows us to do this. It displays a text string in the command window and then waits for the user to provide the requested input. For example,

```
z = input('Enter a temperature value')
```

displays

```
Enter a temperature value
```

in the command window. If the user enters a value such as

```
5
```

the program assigns the value of 5 to the variable **z.** If the input command does not end with a semicolon, the value entered is displayed on the screen.

The same approach can be used to enter a one- or two-dimensional matrix. The user must provide the appropriate brackets and delimiters (commas and semicolons). For example,

```
z = input('Enter values in 2 rows and 3 columms');
```

requests the user to input a matrix such as:

```
[1,2,3;   4,5,6]
```

Since there is not a semicolon after the **input** statement, the following is displayed:

```
z =
    1  2  3
    4  5  6
```

We will give examples using the **input** statement in some of the problem solutions that follow in later chapters.

SUMMARY

In this chapter, we introduced you to the MATLAB environment. In particular, we explored the window structure and solved problems in the command window. The primary data structure in MATLAB is a matrix, which can be a single point (a scalar), a list of values (a vector), or a rectangular grid of values with rows and columns. Values can be entered into a matrix by explicitly listing the values or by loading them from MAT files or ASCII files. In addition, we learned how to save values in both MAT files and ASCII files. We explored the various mathematical operations that are performed in an element-by-element manner. Finally, we learned how to use a script M-file to record the sequence of commands used to solve a MATLAB problem.

MATLAB Summary

This MATLAB summary lists all the special constants, special characters, commands, and functions that were defined in this chapter:

Special Constants	
`ans`	value computed
`clock`	current time
`date`	current date
`eps`	smallest difference recognized
`i`	imaginary number, $\sqrt{-1}$
`Inf`	infinity
`j`	imaginary number, $\sqrt{-1}$
`NaN`	not-a-number
`pi`	mathematical constant, π

Special Characters	
`[ ]`	forms matrices
`( )`	used in statements to group operations and used with a matrix name to identify specific elements
`...`	used to indicate a command continued on the next line
`'`	indicates the transpose of a matrix (apostrophe)
`,`	separates subscripts or matrix elements (comma)
`;`	separates rows in a matrix definition and suppresses output when used in commands
`:`	used to generate matrices and indicates all rows or all columns
`=`	assignment operator—assigns a value to a memory location—not the same as an equality
`%`	indicates a comment in an M-file and used in formatting output
`+`	scalar and element-by-element addition
`-`	scalar and element-by-element subtraction
`*`	scalar multiplication
`.*`	element-by-element multiplication
`/`	scalar division
`./`	element-by-element division
`^`	scalar exponentiation
`.^`	element-by-element exponentiation
`\n`	new line

Functions	
`clc`	clears command screen
`clear`	clears workspace
`disp`	displays strings and variables
`edit`	opens the edit window
`exit`	terminates MATLAB

Functions	
`format`	sets format to default format
`format+`	sets format to plus and minus signs only
`format bank`	sets format to 2 decimal places
`format compact`	sets format to compact form
`format long`	sets format to 14 decimal places
`formal long e`	sets format to 14 exponential places
`format loose`	sets format back to default, noncompact form
`format short`	sets format back to default, four decimal places
`format short e`	sets format to four exponential places
`fprintf`	displays formatted output
`help`	invokes help utility
`input`	allows user to enter values from the keyboard
`iskeyword`	prints a list of keywords
`isvarname`	determines whether a name is a valid variable name
`linspace`	linearly spaced vector function
`load`	loads matrices from a file
`namelengthmax`	displays the number of characters used by MATLAB in a variable name
`quit`	terminates MATLAB
`save`	saves variables in a file
`what`	displays M-files in the current directory
`which`	specifies whether a function is built-in or user-defined
`who`	lists variables in memory
`whos`	lists variables and their sizes

KEY TERMS

array editor
ASCII
assignment operator
colon operator
command history window
command window
comment
current folder window
DAT-file
document window
edit window
element-by-element operations
ellipsis
empty matrix
exponent overflow
exponent underflow
exponential notation
graphics window
M-file
MAT-files
matrix
operation
precedence
precision field
prompt
scalar
scientific notation
script file
transpose operator
up arrow
variable name
vector
width field
workspace window

PROBLEMS

Which of the following are legitimate variable names in MATLAB? Test your answers by trying to assign a value to each name by using, for example,

```
3vars = 3
```

or by using **isvarname**, as in

```
isvarname 3vars
```

Remember, **isvarname** returns a 1 if the name is legal and a 0 if it is not.

1. `3vars`
2. `global`
3. `help`
4. `My_var`
5. `sin`
6. `X+Y`
7. `_input`
8. `input`
9. `tax-rate`
10. `example1.1`
11. `example1_1`
12. Although it is possible to reassign a function name as a variable name, it is not a good idea, so checking to see if a name is also a function name is also recommended. Use **which** to check whether the preceding names are function names, as in

    ```
    which cos
    ```

    Are any of the names in Problems 1 to 11 also MATLAB function names?

Predict the outcome of the following MATLAB calculations. Check your results by entering the calculations into the command window.

13. `1+3/4`
14. `5*6*4/2`
15. `5/2*6*4`
16. `5^2*3`
17. `5^(2*3)`
18. `1+3+5/5+3+1`
19. `(1+3+5)/(5+3+1)`

Create MATLAB code to perform the following calculations. Remember that the square root of a value is equivalent to raising the value to the ½ power. Check your code by entering it into MATLAB.

20. 5^2

21. $\dfrac{5+3}{5 \cdot 6}$

22. $\sqrt{4+6^3}$

23. $9\dfrac{6}{12} + 7 \cdot 5^{3+2}$

24. $1 + 5 \cdot 3/6^2 + 2^{2-4} \cdot 1/5.5$

25. The area of a circle is πr^2. Define r as 5, and then find the area of a circle.

26. The surface area of a sphere is $4\pi r^2$. Find the surface area of a sphere with a radius of 10 ft.

27. The volume of a sphere is $\frac{4}{3}\pi r^3$. Find the volume of a sphere with a radius of 2 ft.

28. The volume of a cylinder is $\pi r^2 h$. Define r as 3 and h as the matrix

```
h = [1,5,12]
```

Find the volume of the cylinders.

29. The area of a triangle is ½ base × height. Define the base as the matrix

```
b = [2,4,6]
```

and the height h as 12, and find the area of the triangles.

30. The volume of a right prism is base area × vertical dimension. Find the volumes of prisms with triangles of Problem 29 as their bases, for a vertical dimension of 10.

31. Generate an evenly spaced vector of values from 1 to 20, in increments of 1. (Use the **`linspace`** command.)

32. Generate a vector of values from zero to 2π in increments of $\pi/100$. (Use the **`linspace`** command.)

33. Generate a vector containing 15 values, evenly spaced between 4 and 20. (Use the **`linspace`** command.)

34. Generate a table of conversions from degrees to radians. The first line should contain the values for 0°, the second line should contain the values for 10°, and so on. The last line should contain the values for 360°.

35. Generate a table of conversions from centimeters to inches. Start the centimeters column at 0 and increment by 2 cm. The last line should contain the value 50 cm.

36. Generate a table of conversions from mi/h to ft/s. The initial value in the mi/h column should be 0 and the final value should be 100. Print 14 values in your table.

37. The general equation for the distance that a free falling body has traveled (neglecting air friction) is

$$d = \tfrac{1}{2}gt^2$$

Assume that $g = 9.8\ \mathrm{m/s^2}$. Generate a table of time versus distance traveled, for time from 0 to 100 s in increments of 10 s. Be sure to use element-by-element operations, and not matrix operations.

38. Newton's law of universal gravitation tells us that the force exerted by one particle on another is

$$F = G\frac{m_1 m_2}{r^2}$$

where the universal gravitational constant is found experimentally to be

$$G = 6.673 \times 10^{-11}\text{N}m^2/\text{kg}^2$$

The mass of each object is m_1 and m_2, respectively, and r is the distance between the two particles. Use Newton's law of universal gravitation to find the force exerted by the Earth on the Moon, assuming that:

the mass of the Earth is approximately 6×10^{24} kg,

the mass of the Moon is approximately 7.4×10^{22} kg, and

the Earth and the Moon are an average of 3.9×10^8 m apart.

39. We know the Earth and the Moon are not always the same distance apart. Find the force the Moon exerts on the Earth for 10 distances between 3.9×10^8 m and 4.0×10^8 m.

MATRIX ANALYSIS

Create the following matrix **A**:

$$\mathbf{A} = \begin{bmatrix} 3.4 & 2.1 & 0.5 & 6.5 & 4.2 \\ 4.2 & 7.7 & 3.4 & 4.5 & 3.9 \\ 8.9 & 8.3 & 1.5 & 3.4 & 3.9 \end{bmatrix}$$

40. Create a matrix **B** by extracting the first column of matrix **A**.
41. Create a matrix **C** by extracting the second row of matrix **A**.
42. Use the colon operator to create a matrix **D** by extracting the first through third columns of matrix **A**.
43. Create a matrix **F** by extracting the values in columns 2, 3, and 4, and combining them into a single column matrix.
44. Create a matrix **G** by extracting the values in columns 2, 3, and 4, and combining them into a single row matrix.

CHAPTER

3 MATLAB Functions

Objectives

After reading this chapter, you should be able to

- use a variety of mathematical and trigonometric functions,
- use statistical functions,
- generate uniform and Gaussian random sequences, and
- write your own MATLAB functions.

ENGINEERING ACHIEVEMENT: WEATHER PREDICTION

Weather satellites provide a great deal of information to meteorologists to use in their predictions of the weather. Large volumes of historical weather data is also analyzed and used to test models for predicting weather. In general, meteorologists can do a reasonably good job of predicting overall weather patterns. However, local weather phenomena, such as tornadoes, water spouts, and microbursts, are still very difficult to predict. Even predicting heavy rainfall or large hail from thunderstorms is often difficult. Although Doppler radar is useful in locating regions within storms that could contain tornadoes or microbursts, the radar detects the events as they occur and thus allows little time for issuing appropriate warnings to populated areas or aircraft passing through the region. Accurate and timely prediction of weather and associated weather phenomena still provides many challenges for engineers and scientists. In this chapter, we present several examples related to analysis of weather phenomena.

3.1 INTRODUCTION TO FUNCTIONS

Arithmetic expressions often require computations other than addition, subtraction, multiplication, division, and exponentiation. For example, many expressions require the use of logarithms, exponentials, and trigonometric functions. MATLAB includes a **built-in** library of these useful functions. For example, if we want to compute the square root of **x** and store the result in **b**, we can use the following commands:

```
x = 9;
b = sqrt(x);
```

If **x** is a matrix, the function will be applied element-by-element to the values in the matrix, as shown in these statements:

```
x = [4, 9, 16];
b = sqrt(x)
b =
    2 3 4
```

All **functions** can be thought of as having three components: a name, input, and output. In this example, the name of the function is **sqrt**, the required input (also called the **argument**) goes inside the parentheses and can be a scalar or a matrix, and the output is a calculated value or values. The output was assigned the variable name **b**.

Some functions require multiple inputs. For example, the remainder function, **rem**, requires two inputs—a dividend and a divisor. We represent this as **rem(x,y)**. The function computes the remainder of the division, and returns that as the output of the function as shown in this example:

```
rem(10, 3)
ans =
     1
```

The **size** function is an example of a function that returns two outputs. This function determines the number of rows and columns in a matrix, and returns the two values as a vector that represents the function output. For example,

```
d = [1,2,3; 4,5,6];
size(d)
ans =
     2 3
```

You can also assign scalar variable names to each of the outputs by representing the left-hand side of the assignment statement as a vector, as in

```
[nrows, ncols] = size(d)
nrows =
        2
ncols =
        3
```

You can create **nested functions**, as shown in this example that computes the square root of the sine of the variable **z**:

```
g = sqrt(sin(z))
```

When one function is used to compute the argument of another function, be sure to enclose the argument of each function in its own set of parentheses. Nesting of functions is also called **composition** of functions.

MATLAB includes extensive help tools, which are especially useful for interpreting function syntax. To use the command-line help function, type **help** in the command window followed by a command or a function name.

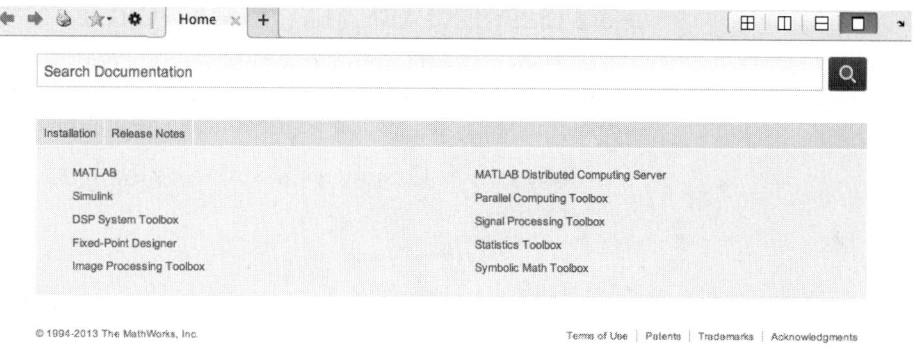

Figure 3.1
The MATLAB Help browser.

For example, to get help on the tangent function, type

```
help tan
```

The following will be displayed:

```
tan Tangent of argument in radians.
tan(X) is the tangent of the elements of X.
See also atan, atan2, tand, atand2d
```

Entering **doc** on the command line will open the Help browser as shown in Figure 3.1. Select MATLAB in this help window. Scroll to the bottom and select functions. At the top right on the page select alphabetical list. You will then get an alphabetical list of all functions, as shown in Figure 3.2

If you enter the word **doc** followed by a command or a function name, MATLAB will display the associated documentation. Experiment with this feature and others in the Help browser.

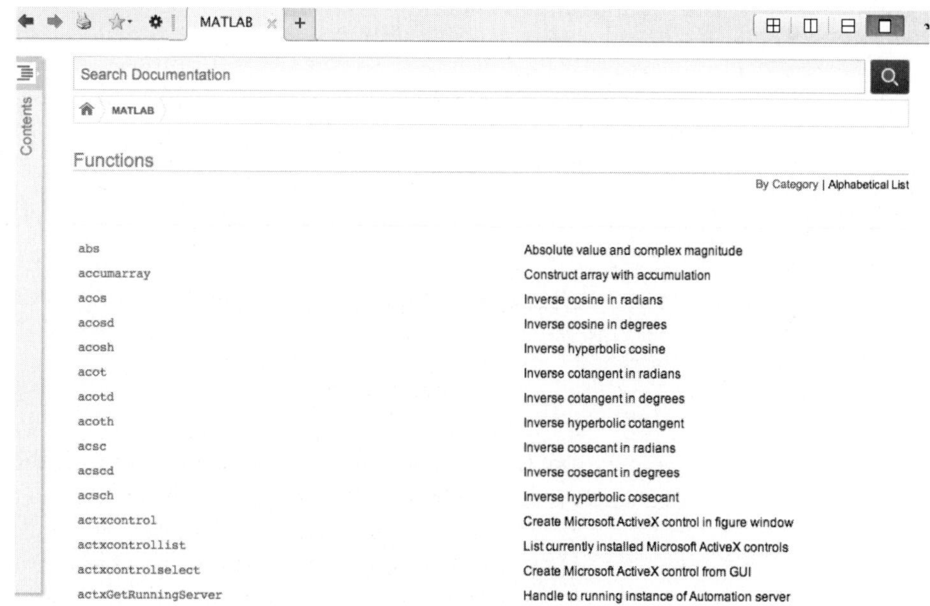

Figure 3.2
Alphabetical listing of MATLAB functions.

3.2 ELEMENTARY MATHEMATICAL FUNCTIONS

The elementary math functions include functions to perform a number of common computations, such as computing the absolute value and the square root of a number. In addition, in this section wc include a group of functions used to perform rounding:

abs(x) Computes the absolute value of x.

```
abs(-3)
ans =
      3
```

sqrt(x) Computes the square root of **x**.

```
sqrt(85)
ans =
      9.2195
```

round(x) Rounds **x** to the nearest integer.

```
round(8.6)
ans =
      9
```

fix(x) Rounds (or truncates) **x** to the nearest integer toward zero.

```
fix(8.6)
ans =
      8
```

floor(x) Rounds **x** to the nearest integer toward −infinity.

```
floor(-8.6)
ans =
      -9
```

ceil(x) Rounds **x** to the nearest integer toward +infinity.

```
ceil(-8.6)
ans =
      -8
```

sign(x) Returns a value of −1 if **x** < 0, and returns a value of 0 if **x** = 0, and returns a value of +1 if **x** > 0.

```
sign(-8)
ans =
      -1
```

rem(x,y) Computes the remainder of **x/y**.

```
rem(25, 4)
ans =
      1
```

exp(x) Computes the value of e^x where e is the base for natural logarithms or approximately 2.718282.

```
exp(10)
ans =
      2.2026e+04
```

`log(x)` Computes ln(**x**), the natural logarithm of **x** to the base *e*.

```
log(10)
ans =
      2.3026
```

`log10(x)` Computes $\log_{10}$(**x**), the common logarithm of **x** to the base 10.

```
log10(10)
ans =
      1
```

`log2(x)` Computes $\log_2$(**x**), the logarithm of **x** to the base 2.

```
log2(8)
ans =
      3
```

Logarithms deserve special mention in this section. As a rule, the function **log** in all computer languages means the **natural logarithm**. Although not the standard in mathematics textbooks, this is the standard in computer programming. This is a common source of errors, especially for new users. If you want the log base 10, you'll need to use the **log10** function. A **log2** function is also included in MATLAB, but logarithms to any other base will need to be computed; there is no general logarithm function that allows the user to input the base.

You can find the syntax for other common mathematical functions by selecting **Help** from the tool bar, and following the **Mathematics** link.

EXAMPLE 3.1

ATMOSPHERIC ANALYSIS

Meteorologists study the atmosphere in an attempt to understand and ultimately predict the weather. Weather prediction is a complicated process, even with the best data. Meteorologists study chemistry, physics, thermodynamics, and geography, in addition to specialized courses about the atmosphere. (See Figure 3.3.) One equation used by meteorologists is the Clausius–Clapeyron equation, which is usually introduced in chemistry classes and examined in more detail in thermodynamics classes.

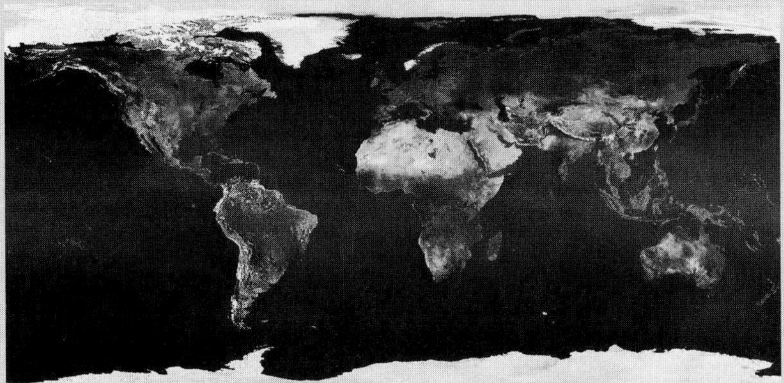

Figure 3.3
View of the Earth's weather from space.

(*continued*)

In meteorology, the Clausius–Clapeyron equation is used to determine the relationship between saturation water-vapor pressure and the atmospheric temperature. The saturation water-vapor pressure can be used to calculate relative humidity, an important component of weather prediction, when the actual partial pressure of water in the air is known.

The Clausius–Clapeyron equation is:

$$\ln(P^\circ/6.11) = \left(\frac{\Delta H_v}{R_{\text{air}}}\right) * \left(\frac{1}{273} - \frac{1}{T}\right)$$

where

P° is the saturation vapor pressure for water, in mbar, at temperature T;
ΔH_v is the latent heat of vaporization for water, 2.45×10^6 J/kg;
R_{air} is the gas constant for moist air, 461 J/kg; and
T is the temperature in degrees K.

It is rare that temperatures on the surface of the Earth are lower than −60°F or higher than 120°F. Use the Clausius–Clapeyron equation to find the saturation vapor pressure for temperatures in this range. Present your results as a table of temperature in Fahrenheit and saturation vapor pressure.

SOLUTION

1. Problem Statement

Find the saturation vapor pressure at temperatures from −60°F to 120°F, using the Clausius–Clapeyron equation.

2. Input/Output Description

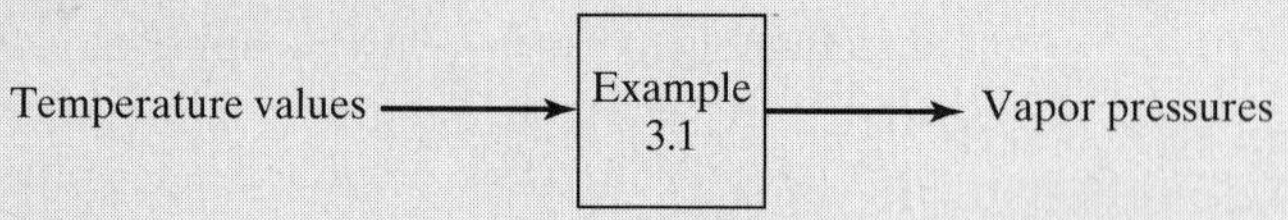

Temperature values → Example 3.1 → Vapor pressures

3. Hand Example

The Clausius–Clapeyron equation requires all the variables to have consistent units. That means that temperature (T) needs to be in degrees K. To change Fahrenheit to Kelvin, use

$$T_k = (T_f + 459.6)/1.8$$

(There are many places to find unit conversions. The Internet is one source, as are science and engineering textbooks.)

Now we need to solve the Clausius–Clapeyron equation for the saturation vapor pressure (P°). We have

$$\ln(P^\circ/6.11) = \left(\frac{\Delta H_v}{R_{\text{air}}}\right) * \left(\frac{1}{273} - \frac{1}{T}\right)$$

$$P^\circ = 6.11 * \exp\left(\left(\frac{\Delta H_v}{R_{\text{air}}}\right) * \left(\frac{1}{273} - \frac{1}{T}\right)\right)$$

Solve for one temperature, for example, let $T = 0°F$:

$$T = (0 + 459.6)/1.8 = 255.3333$$

$$P° = 6.11 * \exp\left(\left(\frac{2.45 \times 10^6}{461}\right) * \left(\frac{1}{273} - \frac{1}{255.3333}\right)\right) = 1.5888 \text{ mbar}$$

4. MATLAB Solution

Create the MATLAB solution in an M-file, and then run it in the command environment:

```
%---------------------------------------------------------------
%  Example 3_1 This program computes the saturation
%  vapor pressure for water at different temperatures.
%
clear, clc
%
%  Define the temperature matrix in F and convert to K.
TF = [-60:10:120];
TK = (TF + 459.6)/1.8;
%
% Define latent heat constant and ideal gas constant.
Delta_H = 2.45e6;
R_air = 461;
%
% Calculate the vapor pressures.
Vapor_Pressure = 6.11*exp((Delta_H/R_air)*(1/273-1./TK));
%
% Display the results in a table.
disp('Temp(k) and Vapor Pressure');
disp([TF',Vapor_Pressure'])
%---------------------------------------------------------------
```

When creating a MATLAB program, it is a good idea to comment liberally. This makes your program easier for others to understand, and may make it easier for you to debug. Notice that most of the lines of code end with a semicolon, which suppresses the output. Therefore, the only information that displays in the command window is the output table.

5. Testing

The output from this program is shown below:

```
Temp(k) and Vapor Pressure
-60.0000  0.0698
-50.0000  0.1252
-40.0000  0.2184
-30.0000  0.3714
-20.0000  0.6163
```

(continued)

```
-10.0000   1.0000
  0        1.5888
  .
  .
  .
100.0000   65.5257
110.0000   88.4608
120.0000   118.1931
```

The hand solution and the MATLAB solution match for $T = 0°F$. The Clausius–Clapeyron equation can be used for more than just humidity problems. By changing the value of ΔH and R, you could generalize the program to any condensing vapor.

3.3 TRIGONOMETRIC FUNCTIONS

The trigonometric functions **sin**, **cos**, and **tan** all assume that angles are represented in radians. The trigonometric functions **sind**, **cosd**, and **tand** all assume that angles are represented in degrees.

In trigonometric calculations, the value of π is often needed, so a constant, **pi,** is built into MATLAB. However, since π cannot be expressed as a floating-point number, the constant **pi** in MATLAB is only an approximation of the mathematical quantity π. Usually, this is not important; however, you may notice some surprising results—for example,

```
sin(pi)
ans =
      1.2246e-16
```

We expected an answer of zero; we got a very small answer, but not zero.

Using the Help browser described in Section 3.1, you can obtain a complete list of trigonometric functions available in MATLAB. Note that there are function for arguments in radians, and functions for arguments in degrees; some of these are listed here:

sin(x) Computes the sine of **x**, where **x** is in radians.

```
sin(pi/2)
ans =
     1
```

sind(x) Computes the sine of **x**, where **x** is in degrees.

```
sind(90)
ans =
     1
```

cos(x) Computes the cosine of **x**, where **x** is in radians.

```
cos(pi)
ans =
    -1
```

tan(x) Computes the tangent of **x**, where **x** is in radians.

```
tan(pi)
ans =
     -1.2246e-16
```

asin(x) Computes the arcsine, or inverse sine, of **x**, where **x** must be between -1 and 1. The function returns an angle in radians between $\pi/2$ and $-\dfrac{\pi}{2}$.

```
asin(-1)
ans =
     -1.5708
```

sinh(x) Computes the hyperbolic sine of **x**, where **x** is in radians.

```
sinh(pi)
ans =
     11.5487
```

EXAMPLE 3.2

COMPUTING DISTANCES USING GPS

The GPS (Global Positioning System) coordinates that specify a location on the Earth are the latitude and longitude values for the position. In this section, we develop an algorithm and MATLAB program to determine the distance between two objects given their latitudes and longitudes. Before developing the programs, we need to briefly discuss latitudes and longitudes and develop an equation to determine the distance between the two points using these coordinates.

Assume that Earth is represented by a sphere with a radius of 3960 miles. A **great circle** is formed by the intersection of this sphere and a plane that passes through the center of the sphere. If the plane does not pass through the center of the sphere, it will be a circle with a smaller circumference and hence is not a great circle. The **prime meridian** is a north–south great circle that passes through Greenwich, just outside London, and through the North Pole. The **equator** is an east–west great circle that is equidistant from the North Pole and the South Pole. Thus, we can define a **rectangular coordinate** system such that the origin is the center of the Earth, the z-axis goes from the center of the Earth through the North Pole, and the x-axis goes from the center of the Earth through the point where the prime meridian intersects the equator. (See Figure 3.4.) The **latitude** is an angular distance, is measured in degrees, and extends northward or southward from the

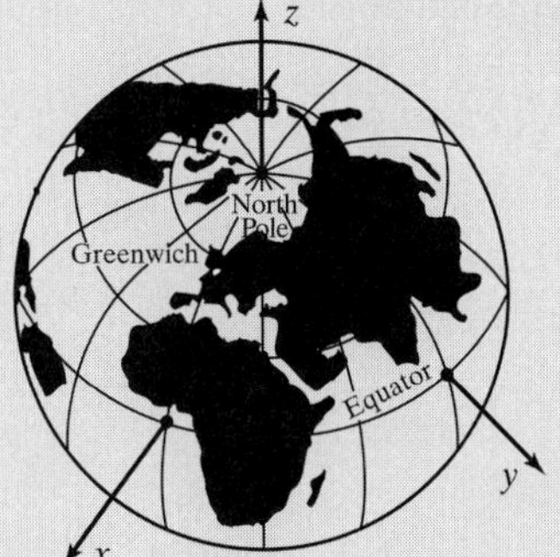

Figure 3.4 Rectangular Coordinate System for the Earth.

(continued)

equator (as in 25 N); and the **longitude** is an angular distance, is measured in degrees, and extends westward or eastward from the prime meridian (as in 120 W).

The **Global Positioning System (GPS)**, originally developed for military use, uses 24 satellites circling the Earth to pinpoint a location on the surface. Each satellite broadcasts a coded radio signal indicating the time and the satellite's exact position 11,000 miles above the Earth. The satellites are equipped with an atomic clock that is accurate to within one second every 70,000 years. A GPS receiver picks up the satellite signal and measures the time between the signal's transmission and its reception. By comparing signals from at least three satellites, the receiver can determine the latitude, longitude, and altitude of its position.

The shortest distance between two points on a sphere is known to be on the arc of the great circle containing them. If we know the angle between vectors from the center of the Earth to the two points defining the arc, we can then estimate the distance as a proportion of the Earth's circumference. To illustrate, suppose that the angle between two vectors from the center of the Earth is 45°. Then the angle is 45/360, or 1/8 of a complete revolution. Hence, the distance between the two points is 1/8 of the Earth's circumference (pi times twice the radius) or 3110 miles.

The best way to compute the shortest distance between two points that are specified in latitude (α) and longitude (β) is through a series of coordinate transformations. Recall that the **spherical coordinates** (ρ, φ, θ) of a point P in a rectangular coordinate system represents the length ρ (rho) of the vector connection the point to the origin, the angle φ (phi) between the positive z-axis and the vector, and the angle θ (theta) between the x-axis and the projection of the vector in the xy-plane. (See Figure 3.5.) We then convert the spherical coordinates to rectangular coordinates (x, y, z). Finally a simple trigonometric equation computes the angle between the points (or vectors) in rectangular coordinates. Once we know the angle between the points, we can then use the technique described in the previous paragraph to find the distance between the two points.

We need to use equations that relate latitude and longitude to spherical coordinates, that convert spherical coordinates to rectangular coordinates, and that compute the angle between two vectors. Figure 3.5 is useful in relating the notation to the following equations:

- Latitude/longitude and spherical coordinates:

$$\alpha = 90^\circ - \varphi, \beta = 360^\circ - \theta$$

- Spherical and rectangular coordinates:

$$\mathrm{x} = \rho \sin\varphi \cos\theta, \mathrm{y} = \rho \sin\varphi \sin\theta, \mathrm{z} = \rho \cos\theta$$

- Angle γ between two vectors a and b:

$$\cos\gamma = \mathrm{a} \cdot \mathrm{b}/(|\mathrm{a}||\mathrm{b}|)$$

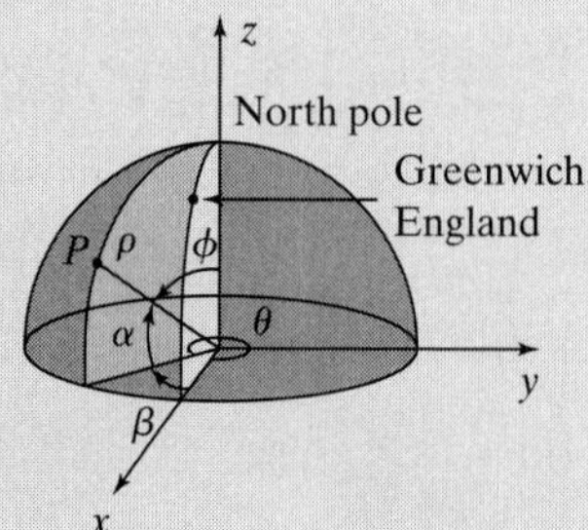

Figure 3.5
Spherical Coordinate System for the Earth.

where a · b is the dot product (defined below) of a and b and where |a| is the length of the vector a (also defined below)

- Dot product of two vectors (xa, ya, za) and (xb, yb, zb) in rectangular coordinates:

$$a \cdot b = xa \cdot xb + ya \cdot yb + za \cdot zb$$

- Length of a vector (xa, ya, za) in rectangular coordinates:

$$\sqrt{(xa^2 + ya^2 + za^2)}$$

- Great circle distance:

$$\begin{aligned} \text{distance} &= (\gamma/(2\pi))(\text{Earth's circumference}) \\ &= (\gamma/(2\pi))(\pi \cdot 2 \text{ radius}) = \gamma \cdot 3960. \end{aligned}$$

Pay close attention to the angle units in these equations. Unless otherwise specified, it is assumed that the angles are measured in radians.

Write a MATLAB program that asks the user to enter the latitude and longitude coordinates for two points in the Northern Hemisphere. Then compute and print the shortest distance between the two points.

1. Problem Statement

Compute the shortest distance between two points in the Northern Hemisphere.

2. Input/Output Description

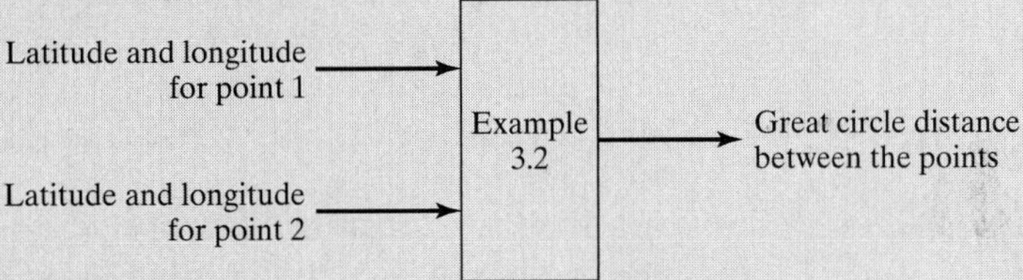

3. Hand Example

For a hand example, we will compute the great circle distance between New York and London. The latitude and longitude of New York is 40.75°N and 74°W, respectively, and the latitude and longitude of London is 51.5°N and 0°W, respectively.

The spherical coordinates for New York are

$$\begin{aligned} \varphi &= (90 - 40.75)^\circ = 49.25(\pi/180) = 0.8596 \text{ radians} \\ \theta &= (360 - 74)^\circ = 286(\pi/180) = 4.9916 \text{ radians} \\ \rho &= 3960. \end{aligned}$$

The rectangle coordinates for New York (to two decimal places) are

$$\begin{aligned} x &= \rho \sin \varphi \cos \theta = 826.90 \\ y &= \rho \sin \varphi \sin \theta = -2883.74 \\ z &= \rho \cos \theta = 2584.93. \end{aligned}$$

Similarly, the rectangular coordinates for London can be computed to be

$$x = 2465.16, y = 0, z = 3099.13.$$

(continued)

The cosine of the angle between the two vectors is equal to the dot product of the two vectors divided by the product of their lengths, or 0.6408. Using an inverse cosine function, γ can be determined to be 0.875 radians. Finally, the distance between New York and London is

$$0.875 \cdot 3960 = 3466 \text{ miles}.$$

4. MATLAB Program

```
%-----------------------------------------------------------------
% Example 3_2 This program determines the distance between
% two points that are specified with latitude and longitude
% values that are in the Northern Hemisphere.
%
clear, clc
%
% Get locations of two points.
lat1 = input('Enter latitude north for point 1: ');
long1 = input('Enter longitude west for point 1: ');
lat2 = input('Enter latitude north for point 2: ');
long2 = input('Enter longitude west for point 2: ');
%
% Convert latitude and longitude to rectangular coordinates.
rho = 3960;
phi = (90 - lat1)*(pi/180);
theta = (360 - long1)*(pi/180);
x1 = rho*sin(phi)*cos(theta);
y1 = rho*sin(phi)*sin(theta);
z1 = rho*cos(phi);
phi = (90 - lat2)*(pi/180);
theta = (360 - long2)*(pi/180);
x2 = rho*sin(phi)*cos(theta);
y2 = rho*sin(phi)*sin(theta);
z2 = rho*cos(phi);
%
% Compute the angle between vectors.
dot = x1*x2 + y1*y2 + z1*z2;
dist1 = sqrt(x1*x1 + y1*y1 + z1*z1);
dist2 = sqrt(x2*x2 + y2*y2 + z2*z2);
gamma = acos(dot/(dist1*dist2));
%
% Compute and print the great circle distance.
display('Great Circle Distance in miles:');
fprintf('%8.0f \n',gamma*rho)
%-----------------------------------------------------------------
```

5. Testing

We start testing with the hand example, which gives the following interaction:

```
Enter latitude north for point 1: 40.75
Enter longitude west for point 1: 74
```

```
Enter latitude north for point 2: 51.5
Enter longitude west for point 2: 0
Great Circle Distance in miles:
   3466
```

This matches our hand example. Try this with some other locations, but remember that you need to choose points that are in the northern latitude and western longitude. The equations need to be slightly modified for points in other parts of the world.

3.4 DATA ANALYSIS FUNCTIONS

Analyzing data is an important part of evaluating test results. MATLAB contains a number of functions that make it easier to evaluate and analyze data. We first present a number of simple analysis functions, and then functions that compute more complicated measures or metrics related to a data set.

3.4.1 Simple Analysis

The following groups of functions are frequently used in evaluating a set of test data: maximum and minimum, mean and median, sums and products, and sorting. We now cover each group separately.

The **max** and **min** functions can be used in a number of ways to determine not only the maximum and minimum values, but also their locations in a matrix. Here are a set of examples to illustrate the various ways to use these functions.

max(x) Returns the largest value in the vector **x**.

```
x = [1,5,3];
max(x)
ans =
     5
```

max(x) Returns a row vector containing the maximum value from each column of the matrix **x**.

```
x = [1,5,3;
     2,4,6];
max(x)
ans =
     2  5  6
```

[a,b] = max(x) Returns a vector containing the largest value in a vector **x** and its location in the vector **x**.

```
x = [1,5,3];
[a,b] = max(x)
a =
   5
b =
   2
```

[a,b] = max(x) Returns a row vector **a** containing the maximum element from each column of the matrix **x**, and returns a row vector **b**

containing location of the maximum in each column of the matrix **x**.

```
x = [1,5,3;
     2,4,6];
[a,b] = max(x)
a =
    2  5  6
b =
    2  1  2
```

max(x,y) Returns a matrix the same size as **x** and **y**. (Both **x** and **y** must have the same number of rows and columns.) Each element in the resulting matrix contains the maximum value from the corresponding positions in **x** and **y**.

```
x = [ 1,5,3; 2,4,6];
y = [10,2,4; 1,8,7];
max(x,y)
ans =
      10  5  4
       2  8  7
```

min(x) Returns the smallest value in the vector **x**.

```
x = [1,5,3];
min(x)
ans =
      1
```

min(x) Returns a row vector containing the minimum value from each column of the matrix **x**.

```
x = [1,5,3;
     2,4,6];
min(x)
ans =
      1  4  3
```

[a,b] = min(x) Returns a vector containing the smallest value in a vector **x** and its location in the vector **x**.

```
x = [1,5,3];
[a,b] = min(x)
a =
    1
b =
    1
```

[a,b] = min(x) Returns a row vector **a** containing the minimum element from each column of the matrix **x**, and returns a row vector **b** containing location of the minimum in each column of the matrix **x**.

```
x = [1,5,3;
     2,4,6];
```

```
[a,b] = min(x)
a =
   1  4  3
b =
   1  2  1
```

`min(x,y)` Returns a matrix the same size as **x** and **y**. (Both **x** and **y** must have the same number of rows and columns.) Each element in the resulting matrix contains the minimum value from the corresponding positions in **x** and **y**.

```
x = [ 1,5,3; 2,4,6];
y = [10,2,4; 1,8,7];
min(x,y)
ans =
      1  2  3
      1  4  6
```

The **mean** of a group of values is the average of the values. The Greek symbol μ (mu) represents the value of the mean in many mathematics and engineering applications:

$$\mu = \frac{\sum_{k=1}^{N} x_k}{N}$$

$$\sum_{k=1}^{N} x_k = x_1 + x_2 + x_3 + \ldots + x_N$$

In simple terms, to find the mean, just add up all the values and divide by the total. The **median** is the value in the middle of the group, assuming that the values are sorted. If there is an odd number of values, the median is the value in the middle position. If there is an even number of values, then the median is the mean of the two middle values. The functions for computing the mean and the median are as follows:

`mean(x)` Computes the mean value (or average value) of a vector **x**.

```
x = [1,5,3];
mean(x)
ans =
      3
```

`mean(x)` Returns a row vector containing the mean value from each column of a matrix **x**.

```
x = [1,5,3;
      2,4,6];
mean(x)
ans =
      1.5000   4.5000   4.5000
```

`median(x)` Finds the median of the elements of a vector **x**.

```
x = [1,5,3];
median(x)
ans =
      3
```

median(x) Returns a row vector containing the median of each column of a matrix **x**.

```
x = [1,5,3;
     2,4,6];
median(x)
ans =
     1.5000    4.5000    4.5000
```

MATLAB also contains functions for computing the sums and products of vectors (or of the columns in a matrix) and functions for computing the cumulative sums and products of vectors (or the elements of a matrix):

sum(x) Computes the sum of the elements of the vector **x**.

```
x = [1,5,3];
sum(x)
ans =
     9
```

sum(x) Computes a row vector containing the sum of the elements from each column of the matrix **x**.

```
x = [1,5,3;
     2,4,6];
sum(x)
ans =
     3 9 9
```

prod(x) Computes the product of the elements of the vector **x**.

```
x = [1,5,3];
prod(x)
ans =
     15
```

prod(x) Computes a row vector containing the product of the elements from each column of the matrix **x**.

```
x = [1,5,3;
     2,4,6];
prod(x)
ans =
     2 20 18
```

cumsum(x) Computes a vector of the same size as **x** containing the cumulative sums of the elements of **x** from each column of the matrix **x**.

```
x = [1,5,3;
     2,4,6];
cumsum(x)
ans =
     1  5  3
     3  9  9
```

cumprod(x) Computes a vector of the same size as **x** containing the cumulative products of the elements of **x** from each column of the matrix **x**.

```
x = [1,5,3;
     2,4,6];
cumprod(x)
ans =
     1   5   3
     2  20  18
```

The **sort** command arranges the values of a vector **x** into ascending order. If **x** is a matrix, the command sorts each column into ascending order:

sort(x) Sorts the elements of the vector **x**.

```
x = [1,5,3];
sort(x)
ans =
     1  3  5
```

sort(x) Computes a matrix containing the sorted elements from each column of the matrix **x**.

```
x = [1,5,3;
     2,4,6];
sort(x)
ans =
     1  4  3
     2  5  6
```

HINT

All of the functions in this section work on the columns in two-dimensional matrices. If your data analysis requires you to evaluate data in rows, the data must be transposed—in other words, the rows must become columns and the columns must become rows. The transpose operator is a single quote ('). For example, if you want to find the maximum value in each row of matrix **x**, use this command:

```
x = [1,5,3;
     2,4,6];
max(x')
ans =
     5  6
```

MATLAB offers two functions that allow us to determine the size of a matrix: **size** and **length**:

size(x) Determines the number of rows and columns in the matrix **x**.

```
x = [1,5,3;
     2,4,6];
```

```
size(x)
ans =
     2   3
```

[a,b] = size(x) Determines the number of rows in **x** and assigns that value to **a**, and determines the number of columns in **x** and assigns that value to **b**.

```
x = [1,5,3;
     2,4,6];
[a,b] = size(x)
a =
   2
b =
   3
```

length(x) Determines the largest dimension of the matrix **x**; this is also the value of **max(size(x))**.

```
x = [1,5,3;
     2,4,6];
length(x)
ans =
     3
```

EXAMPLE 3.3

PEAK WIND SPEEDS FROM MOUNT WASHINGTON OBSERVATORY

Rising to 6288 ft., New Hampshire's Mount Washington is not among the world's highest mountains. Nevertheless, it lays claim to being the "Home of the World's Worst Weather." Extreme cold, freezing fog, snow, and wind . . . especially wind . . . are commonplace here, as shown in Figure 3.6.

Figure 3.6 Mount Washington Observatory.

Table 3.1 Peak Monthly Peak Wind Speed (mph)

	2005	2006	2007	2008
January	113	142	121	110
February	117	137	127	118
March	117	119	117	145
April	95	110	156	118
May	90	88	113	93
June	58	86	89	84
July	61	99	72	88
August	82	97	94	85
September	129	96	95	93
October	136	158	110	97
November	110	92	110	110
December	104	102	124	129

On April 10, 1934, weather maps showed a weak storm developing well to the west of New Hampshire, another system off the coast of North Carolina, and a high pressure ridge building over eastern Canada and the North Atlantic. The high pressure ridge strengthened the following day, blocking the systems approaching from the west and south, causing them to converge on Mount Washington and the research station at its summit. As night came, winds were already well over 100 mph and the research team at the Observatory prepared for superhurricane conditions. April 12 brought even more strengthening winds with gusts frequently over 220 mph, until 1:21 p.m., when the peak of 231 mph was recorded. It remains today the highest surface wind speed ever officially recorded on Earth.

Mount Washington Observatory collects extensive weather data, some of which can be obtained from their Web site, www.mountwashington.org. Analyzing large amounts of such data can be confusing, so it is a good idea to start with a smaller data set, develop an approach that works, and then apply it to larger data sets. Table 3.1 provides one such set of monthly peak wind data from four successive years, recorded by the Observatory. Enter the data in an Excel file named **`WindData`**, such that each row of the file represents one month and each column represents a year. The Excel file should contain only the data shown in Table 3.1, not the text information.

1. Problem Statement

Using the data in the file **`WindData.xls`**, find the average peak wind speed by month, the average peak wind speed by year, and the month and year that recorded the highest peak wind speed.

2. Input/Output Description

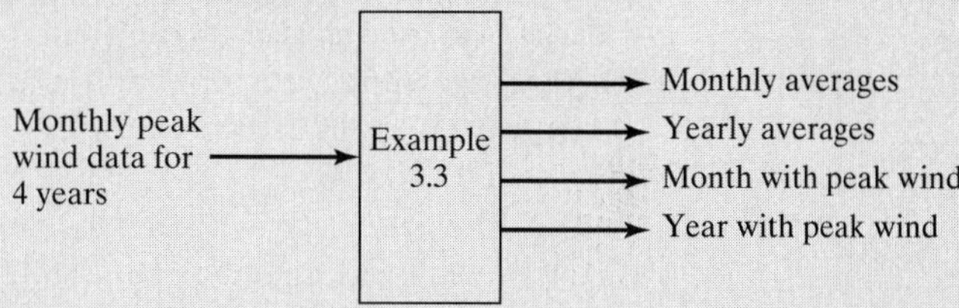

(continued)

3. Hand Example

The data set is small enough to allow us to find some of the results we expect to obtain later with our MATLAB solution. For example, the average of the peak wind speeds for the month of January is

$$(113 + 142 + 121 + 110)/4 = 121.5 \text{ mph}$$

The average peak wind speed for the year 2005 is

$$(113 + 117 + 117 + 95 + 90 + 58 + 61 + 82 + 129 + 136 + 110 + 104) = 101 \text{ mph}$$

Finally, we can see by inspecting the data in the table that the highest overall peak wind speed was 158 mph, occurring in October, 2006. This hand example will help us develop the MATLAB solution for the data set and ensure it is working correctly for this set as well as larger ones.

4. MATLAB Solution

First, we will put the data file into the MATLAB workspace as a matrix. Since it is an Excel spreadsheet, the easiest approach is to use the Import Data selection from the Home tab. When you select Import Data, a window will appear with the files that are in your current directory. Choose the **WindData** spreadsheet and MATLAB will display the values as shown in Figure 3.7. At the top middle of the screen choose a matrix representation for the data, and on the top right of the screen choose Import Data from Import Selection. We now write the script to solve the problem:

```
%-----------------------------------------------------------
% Example 3_3 This program calculates statistics
% from a set of data from Mount Washington.
%
clc
%
% Use Import Data to copy the wind data from
% an excel spreadsheet into the variable WindData.
% Then copy the data into wd.
wd = WindData;
%
% Find the monthly averages (stored in the rows).
disp('Monthly Averages')
disp(sum(wd')/4)
%
% Find the yearly averages (stored in the columns).
disp('Yearly Averages')
disp(sum(wd)/12)
%
% Find the month in which the peak wind occurred
disp('Peak Wind and Corresponding Month')
[peak,month] = max(max(wd'))
%
% Find the year in which the peak wind occurred
disp('Peak Wind and Corresponding Year')
[peak,year] = max(max(wd))
%-----------------------------------------------------------
```

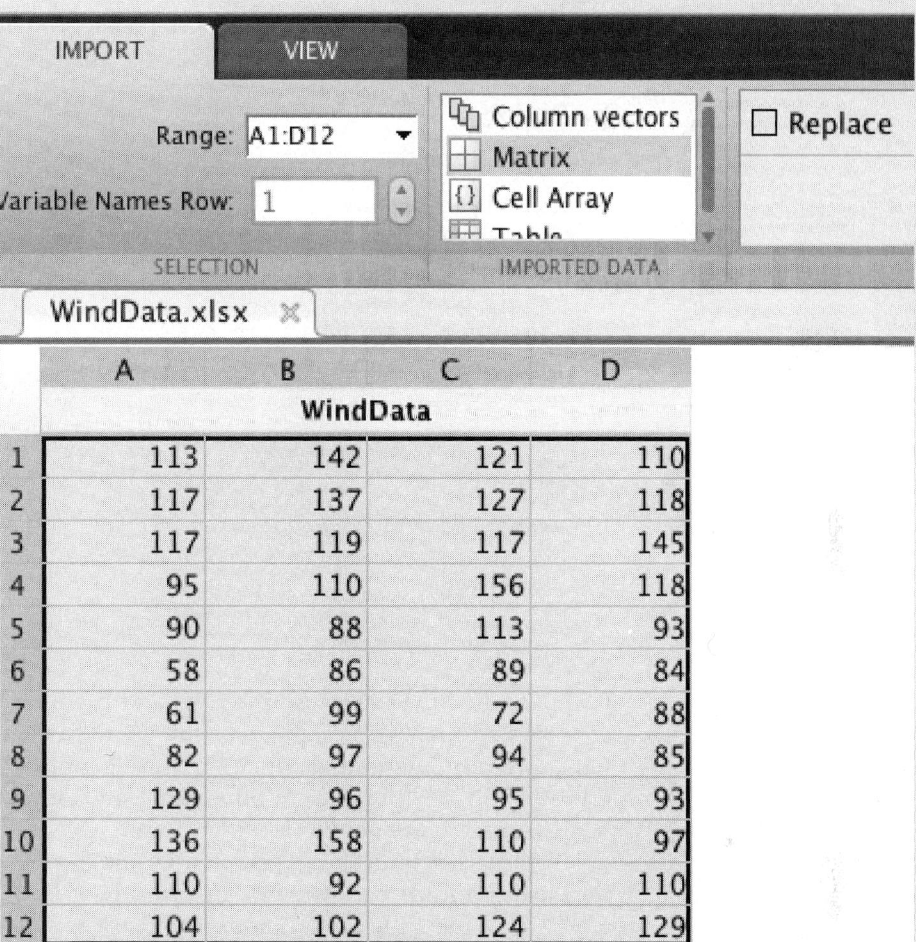

	A	B	C	D
1	113	142	121	110
2	117	137	127	118
3	117	119	117	145
4	95	110	156	118
5	90	88	113	93
6	58	86	89	84
7	61	99	72	88
8	82	97	94	85
9	129	96	95	93
10	136	158	110	97
11	110	92	110	110
12	104	102	124	129

Figure 3.7
MATLAB Import Data.

Notice that the code did not start with our usual `clear, clc` commands, because that would clear the workspace, effectively deleting the **`WindData`** variable. Next, we rename **`WindData`** to **`wd`**, to make the name shorter. We will be using this variable frequently, so it is a good idea to make it short to minimize the chance of errors caused by mistyping.

Next, the matrix **`wd`** is transposed, so that the data for each month is in a column instead of a row. That allows us to use the **`sum`** command to add up all the wind values for the month. Dividing by 4 will calculate the averages for each month. Similarly, we can add up all the monthly totals to get the total for the year and divide by 12 to get the averages for each year.

The transpose of the matrix is used for determining the overall peak wind value and the month in which it occurred. Understanding this command line is easier if we break the command into two commands. The first part of the command, **`max(wd')`**, computes a vector with the maximum value for each month. Thus, **`max(max(wd;))`** will determine the maximum of all 12 months and will also determine which month. A similar operation with **`wd`**, instead of **`wd'`**, will allow us to determine the year in which the peak wind occurred. We can see already from this

(continued)

output that the maximum speed of 158 mph occurred in the tenth month (October) and second year (2006).

5. Testing

The output from this program is shown in below:

```
Monthly Averages
 121.5000  124.7500  124.5000  119.7500   96.0000   79.2500
  80.0000   89.5000  103.2500  125.2500  105.5000  114.7500
Yearly Averages
 101.0000  110.5000  110.6667  105.8333
Peak Wind and Corresponding Month
peak =
      158
month =
       10
Peak Wind and Corresponding Year
peak =
      158
year =
      2
```

Compare the MATLAB output to the hand example computation done earlier to confirm that the averages, the maximum, and the corresponding month and year were correctly determined. Once you have confirmed that the M-file works properly, you can use it to analyze other data sets, especially much larger ones where hand examples may not be easy to perform.

There are several things you should notice after executing this M-file. In the workspace window, both **WindData** and **wd** are listed. All of the variables created when the M-file **Example3_3** is executed are available for additional calculations, to be entered in the command window, if desired. Also notice that the original Excel file, **WindData.xls**, is still in the current directory. Finally, notice that the command history window only reflects commands issued from the command window. It does not show commands executed from an M-file.

3.4.2 Variance and Standard Deviation

Two of the most important statistical measurements of a set of data are the **variance** and the **standard deviation**. Before we give their mathematical definitions, it is useful to develop an intuitive understanding of these values. Consider the values of vectors **data_1** and **data_2**, plotted in Figures 3.8(a) and 3.8(b).

If we attempt to draw a line through the middle of the values in the plots, this line would be at approximately 3.0 in both plots. Thus, we would assume that both vectors have approximately the same mean value of 3.0. However, the data in the two vectors clearly have some distinguishing characteristics. The data values in **data_2** vary more, or deviate more, from the mean. Thus, measures of variance and deviation for the values in **data_2** will be greater than measures of variance and deviation for **data_1**. An intuitive understanding of variance (or deviation)

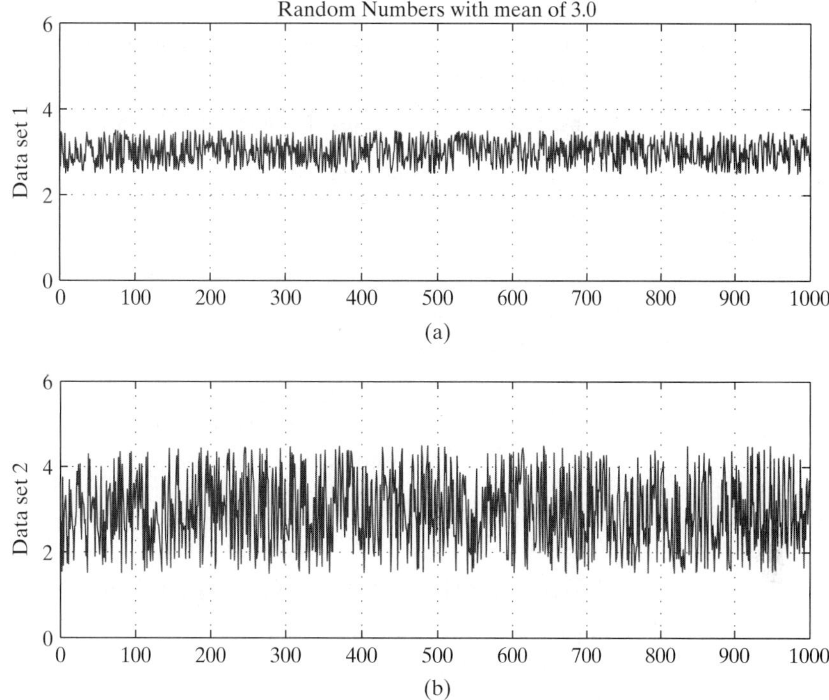

Figure 3.8
(a) Random data (`data_1`) with a mean of 3.0.
(b) Random data (`data_2`) with a mean of 3.0.

relates to the variance of the values from the mean. The larger the variance, the further the values fluctuate from the mean value.

Mathematically, the variance σ^2 of a set of data values (which we will assume are stored in a vector **x**) can be computed with the following equation:

$$\sigma^2 = \frac{\sum_{k=1}^{N} (x_k - \mu)^2}{N - 1}.$$

This equation is a bit intimidating at first, but if you look at it closely it seems much simpler. The term $x_k - \mu$ is the difference, or deviation, of x_k from the mean. This value is squared so that we will always have a positive value. We then add together the squared deviations for all of the data points. This sum is then divided by $N - 1$, which approximates an average. (The equation for variance sometimes uses a denominator of N, but the form here has statistical properties that make $N - 1$ generally more desirable.) Thus, the variance is the average squared deviation of the data from the mean.

The standard deviation is defined as the square root of the variance, or

$$\sigma = \sqrt{\sigma^2}$$

where σ is the Greek symbol sigma. In a normal (Gaussian) distribution of a large amount of data, approximately 68 percent of the data falls within one sigma variation of the mean ($\pm$one sigma). If you extend the range to a two-sigma variation ($\pm$two sigma), approximately 95 percent of the data should fall inside these bounds, and if you go out to three sigma, over 99 percent of the data should fall in this range. (See Figure 3.9.)

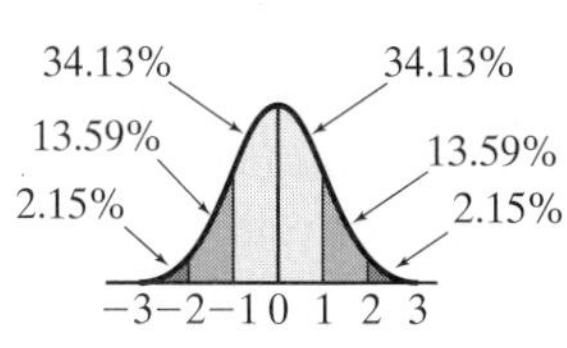

Figure 3.9
Normal distribution of data.

The function for calculating standard deviation is shown below. (MATLAB does not have a built-in function for variance, but to compute the variance we just square the standard deviation.)

std(x) Computes the standard deviation of the elements of a vector **x**.

```
x = [1,5,3];
std(x)
ans =
     2
```

std(x) Returns a row vector containing the standard deviation of each column of a matrix **x**.

```
x = [1,5,3;
     2,4,6];
std(x)
ans =
     0.7071   0.7071   2.1213
```

3.5 RANDOM NUMBER GENERATING FUNCTIONS

There are many engineering problems that require the use of random numbers in the development of a solution. In some cases, the random numbers are used to develop a simulation of a complex problem. The simulation can be tested over and over to analyze the results, and each test represents a repetition of the experiment. We also use random numbers to approximate noise sequences. For example, the static we hear on a radio is a noise sequence. If we are testing a program that uses an input data file that represents a radio signal, we may want to generate noise and add it to a speech signal or music signal to provide a more realistic sound.

3.5.1 Uniform Random Numbers

Random numbers are not defined by an equation; instead, they can be characterized by a distribution of values. For example, random numbers that are equally likely to be any value between an upper and lower limit are called **uniform random numbers**.

The **rand** function in MATLAB generates random numbers uniformly distributed over the interval 0 to 1. If the **rand** function has a single argument **n**, then it generates a matrix with **n** rows and **n** columns filled with random numbers between 0 and 1. If the **rand** function has two arguments **m** and **n**, then it generates a matrix with **m** rows and **n** columns filled with random numbers between 0 and 1.

A seed value is used to initialize a random sequence of values. When you first launch MATLAB, this seed value is given a default value. Each time that you generate random numbers, the seed value changes so that you continue to get different values. If you want to get the same random numbers, you can reset the seed to the default value. There may be times that you want to be able to restart a random number sequence at another point other than the default value. You can do that with statements that store the seed in a variable, and then at a later point in your program, you can reset the random seed using that variable.

We now give the general format of these functions and some examples.

rng('default') Sets the value of the seed to a default seed. This allows you to restart the generation of random numbers so that you can create the same sequence of random numbers.

rng Represents the current value of the seed that can be stored in another variable. To save this value in a variable, and then reset the seed at a later point to this value, use the following statements:

```
s = rng;
...
rng(s);
```

rand(n) Returns a matrix with **n** rows and **n** columns. Each value in the matrix is a uniform random number between 0 and 1.

```
rng('default');
rand(2)
ans =
      0.8147   0.1270
      0.9058   0.9134
```

rand(m,n) Returns a matrix with **m** rows and **n** columns. Each value in the matrix is a uniform random number between 0 and 1.

```
rng('default');
rand(3,2)
ans =
      0.8147   0.9134
      0.9058   0.6324
      0.1270   0.0975
```

You don't need to keep repeating the statement to reset the random number seed to the default value, but if you don't repeat it, you will get a different set of random numbers. Try entering the statements without resetting the seed value to see that you do indeed get different numbers between 0 and 1.

Random sequences with values that range between values other than 0 and 1 are often needed. Suppose we want to generate values between −5 and +5. First, we generate random numbers between 0 and 1, and store them in the matrix **r**:

```
r = rand(100,1);
```

Because the difference between −5 and +5 is 10, we know that we want our random numbers to vary over a range of 10, so we will need to multiply everything by 10:

```
r = r*10;
```

Now our random numbers vary from 0 to 10. If we add the lower bound (−5) to our matrix, with this command

```
r = r-5;
```

the result will be random numbers varying from −5 and +5. We can generalize these results with the equation

$$x = (b - a) \cdot r + a$$

where

a is the lower bound,
b is the upper bound, and
r is a set of random numbers.

We also frequently need to generate random integers between two other integers. The function **randi** generates integers. It can be used to generate integers between 1 and a maximum integer, or between a minimum integer and a maximum integer. The random number seed applies to random integers just as it did with random numbers between 0 and 1.

We now give the general format of this function and some examples.

randi(imax,n) Returns a matrix with **n** rows and **n** columns. Each value in the matrix is a uniform random integer between 1 and **imax**.

```
rng('default');
randi(50,2)
ans =
        41   7
        46  46
```

randi(imax,m,n) Returns a matrix with **m** rows and **n** columns. Each value in the matrix is a uniform random number between 1 and **imax**.

```
rng('default');
randi(20,3,2)
ans =
        17  19
        19  13
        3    2
```

randi(imax) Returns a scalar value between 1 and **imax**.

```
rng('default');
randi(10)
ans =
        9
```

randi([imin,imax],n) Returns a matrix with **n** rows and **n** columns. Each value in the matrix is a uniform random number between **imin** and **imax.**

```
rng('default');
randi([-10,-1],3,2)
ans =
        -2  -1
        -1  -4
        -9  -1
```

3.5.2 Gaussian (Normal) Random Numbers

When we generate a random number sequence with a uniform distribution, all values are equally likely to occur. However, we sometimes need to generate random numbers using distributions in which some values are more likely to be generated than others. For example, suppose that a random number sequence represents outdoor temperature measurements taken over a period of time. We would find that the temperature measurements have some variation, but typically are not equally likely. For example, we might find that the values vary only a few degrees, although larger changes would occasionally occur because of storms, cloud shadows, and day-to-night changes.

Random number sequences that have some values that are more likely to occur than others are often modeled as **Gaussian random numbers** (also called **normal random numbers**). An example of a set of values with a Gaussian distribution is shown in Figure 3.9. Although a uniform random variable has specific upper and lower bounds, a Gaussian random variable is not defined in terms of upper and lower bounds; it is defined in terms of the mean value and the variance, or standard deviation, of the values. For Gaussian random numbers, approximately 68 percent of the values will fall within one standard deviation, 95 percent within two standard deviations, and 99 percent within three standard deviations from the mean.

MATLAB will generate Gaussian values with a mean of 0 and a variance of 1.0 if we specify a normal distribution. The functions already presented to specify the default value of the seed, or to save and reset the seed, work independently of whether the random numbers are uniform or Gaussian. The functions for generating Gaussian values are as follows:

`randn(n)` Returns a matrix with **`n`** rows and **`n`** columns. Each value in the matrix is a Gaussian, or normal, random number with a mean of 0 and a variance of 1.

```
rng('default')
randn(2)
ans =
       0.5377  -2.2588
       1.8339   0.8622
```

`rand(m,n)` Returns a matrix with **`m`** rows and **`n`** columns. Each value in the matrix is a Gaussian, or normal, random number between 0 and 1.

```
rng('default')
randn(3,2)
ans =
        0.5377   0.8622
        1.8339   0.3188
       -2.2588  -1.3077
```

To modify Gaussian values with a mean of 0 and a variance of 1 to another Gaussian distribution, multiply the values by the standard deviation of the desired distribution, and add the mean of the desired distribution. Thus, if r is a random number sequence with a mean of 0 and a variance of 1.0, the following equation will generate a new random number with a standard deviation of a and a mean of b:

$$x = a \cdot r + b$$

For example, to create a sequence of 500 Gaussian random variables with a standard deviation of 2.5 and a mean of 3, use

```
x = randn(1,500)*2.5 + 3;
```

EXAMPLE 3.4

PEAK WIND STANDARD DEVIATION FROM MOUNT WASHINGTON OBSERVATORY

Climatologists examine weather data over long periods of time, trying to find patterns. Weather data have been kept reliably in the United States since the 1850s; however, most reporting stations have only been in place since the 1930s and 1940s.

(continued)

Climatologists perform statistical analyses on the data they collect to detect recurring patterns, trends, and anomalies. Although the data in **WindData.xls** (discussed in Example 3.3) only represent one location for four years, we can use the data to practice statistical calculations. Find the standard deviation for each month and for each year, and then find the standard deviation for the entire data set.

1. Problem Statement

Find the standard deviation for each month, each year, and for the entire data set.

2. Input/Output Description

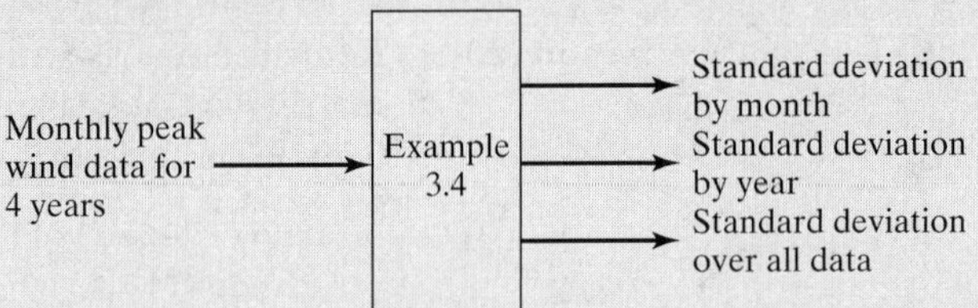

3. Hand Example

The standard deviation is found with the equation from page 79. Using only the month of January, first calculate the sum of the squares of the difference between the mean value of 121.5 and the actual value:

$$(113 - 121.5)^2 + (142 - 121.5)^2 + (121 - 121.5)^2 + (110 - 121.5)^2 = 625$$

Divide this sum by the number of data points minus 1:

$$625/(4 - 1) = 208.3333$$

Finally, take the square root to get 14.4338 mph.

4. MATLAB Solution

```
%-------------------------------------------------------------
% Example 3_4 This program calculates the mean
% peak wind speed by month and by year, and finds the
% standard deviation of the peak speeds by month,
% by year, and for all four years.
%
clc
% Use Import Data to copy the wind data from
% an excel spreadsheet into the variable WindData.
% Then copy the data into wd.
wd = WindData;
%
% Find the standard deviation for each month
disp('Monthly Standard Deviation')
monthly = std(wd');
disp(monthly(1:6))
disp(monthly(7:12))
%
% Find the standard deviation for each year
disp('Yearly Standard Deviation')
```

```
disp(std(wd))
%
% Find the standard deviation for all months and years
disp('Overall Standard Deviation')
disp(std(wd(:)))
%----------------------------------------------------------------
```

The transpose operator is used in this example to find the mean values by month. When we use the matrix again without the transpose, it finds the mean values by year. Finally, recall that

```
wd(:)
```

converts the two-dimensional matrix **wd** into a one-dimensional vector, thus making it possible to find the standard deviation of all the entire data set in one command.

5. Testing

When we execute this M-file, the following results appear in the command window:

```
Monthly Standard Deviation
14.4338   9.3229    13.6991   25.9792   11.5181  14.3149
16.8325   7.1414    17.2119   27.1953    9.0000  13.7447
Yearly Standard Deviation
24.7203   23.4889   21.6179   19.1588
Overall Standard Deviation
21.9894
```

First, check the results visually to make sure they make sense. Notice that the mean values and that the standard deviations vary much more from month-to-month than they do from year-to-year, which is what we would expect. Also compare the MATLAB output to the hand example done earlier to confirm that the standard deviations were correctly determined. As before, once you have confirmed the M-file works properly, you can use it to analyze other data sets.

EXAMPLE 3.5

FLIGHT SIMULATOR

Computer simulations are used to generate situations that model or emulate a real world situation. Some computer simulations are written to play games such as checkers, poker, and chess. To play the game, you indicate your move, and the computer will select an appropriate response. Other animated games use computer graphics to develop an interaction as you use the keys or a mouse to play the game. In more sophisticated computer simulations, such as those in a flight simulator, the computer not only responds to the input from the user, but also generates values, such as temperatures, wind speeds, and the locations of other aircraft. The simulators also model emergencies that occur during the flight of an aircraft. If all of this information generated by the computer were always the same set of information,

(*continued*)

the value of the simulator would be greatly reduced. It is important that there be randomness to the generation of the data. Simulations that use random numbers to generate values that model events are called **Monte Carlo simulations**.

Write a program to generate a random-number sequence to simulate one hour of wind speed data that are updated every 10 seconds. Assume that the wind speed will be modeled as a uniform distribution, random number that varies between a lower limit and an upper limit. Let the lower limit be 5 mph and the upper limit be 10 mph. Save the data to an ASCII file named **`windspd.dat`**.

SOLUTION

1. Problem Statement

Generate one hour of wind speed data using a lower limit of 5 mph and an upper limit of 10 mph.

2. Input/Output Description

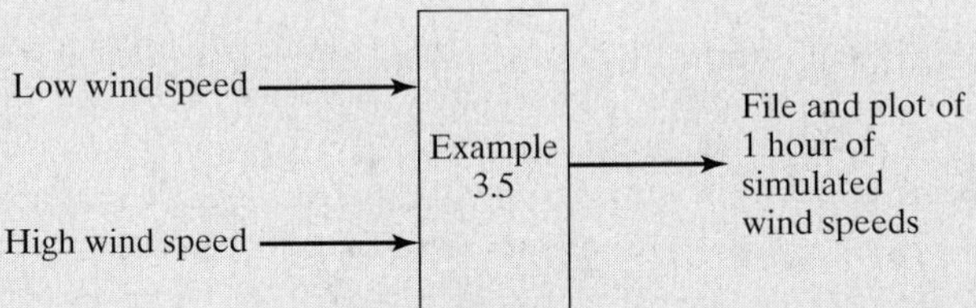

3. Hand Example

This simulation uses MATLAB's random number generator to generate numbers between 0 and 1. We then modify these values to be between a lower limit (5) and an upper limit (10). Using the equation developed in Section 3.5.1,

$$x = (b - a) \cdot r + a$$

we first multiply the range (10-5) times the random numbers, and then add the lower bound (5). Hence, the value 0.1 would be converted to this value:

$$x = (10 - 5) \cdot 0.1 + 5 = 5.5$$

4. MATLAB Solution

```
%--------------------------------------------------------------
% Example 3_5 This program generates and plots
% one hour of simulated wind speeds.
%
clear, clc
%
% Specify high and low speeds.
low_speed = input('Enter low speed: ');
high_speed = input('Enter high speed: ');
%
% Define the time matrix.
t = 0:10:3600;
%
% Convert time to hours.
```

```
t = t/3600;
%
% Determine how many time values are needed.
num = length(t);
%
% Calculate the speed values.
speed = (high_speed - low_speed)*rand(1,num) + low_speed;
%
% Create a table.
table = [t',speed'];
save windspd.dat table -ascii
subplot(2,1,1),plot(t,speed),title('Simulated Wind Speed'),
   xlabel('Time, s'),ylabel('Wind Speed, mph'),grid
%-----------------------------------------------------------
```

Notice that we used the **length** command to determine how many elements were in the **t** matrix; this function returns the maximum dimension of a matrix, and since **t** has 1 row and 361 columns, the result was 361. Also note that we did not set the random number seed to the default value, so each time that the program is run it will generate a different set of simulated wind speeds.

5. Testing

We can see from Figure 3.10 that the wind speed does indeed vary between 5 mph and 10 mph. Since we did not specify a random number seed, this program will output a new set of random wind speeds each time that it is run.

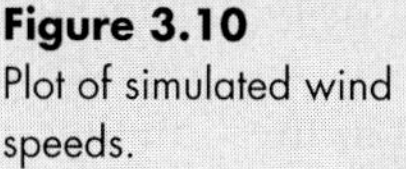

Figure 3.10
Plot of simulated wind speeds.

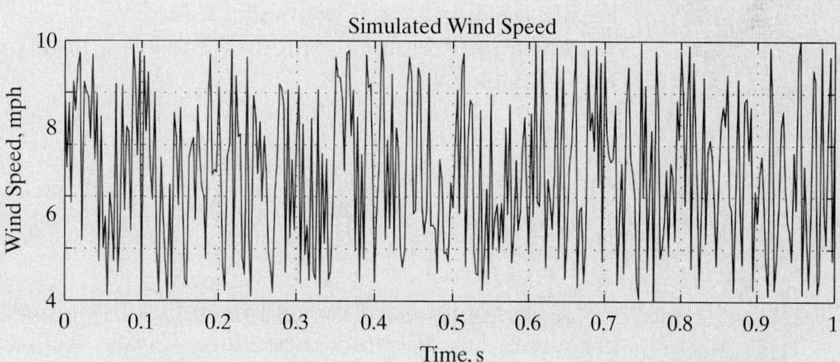

3.6 USER-DEFINED FUNCTIONS

The MATLAB programming language is built around functions. A function is simply a piece of computer code that accepts an input argument from the user and provides output to the program. Functions allow us to program efficiently, since we do not need to rewrite the computer code for calculations that are performed frequently. For example, most computer programs contain a function that calculates the sine of a number. In MATLAB, **sin** is the function name used to call up a series of commands that perform the necessary calculations. The user needs to provide an angle, and MATLAB returns a result. It is not necessary for the user to even know that MATLAB uses an approximation to an infinite series to find the value of **sin**.

We have already explored many of MATLAB's **built-in functions**, but you may wish to define your own functions that are used commonly in your programming. **User-defined functions** are stored as M-files, and can be accessed by MATLAB if they are in the current directory.

3.6.1 Syntax

User defined MATLAB functions are written in M-files. Create a new function M-file the same way a script M-file is created: select New Script from the Home tab. To explain how M-file functions work, we will use several examples.

Here is a simple function to begin:

```
function s=f(x)
% This function adds 3 to each value in an array.
s = x + 3;
```

These lines of code define a function called **f**. Notice that the first line starts with the word **function**. This is a requirement for all user-defined functions. Next, an output variable that we have named **s** is set equal to the function name, with the input arguments enclosed in parentheses **(x)**. A comment line follows that will be displayed if a user types:

```
help f
```

once the function has been saved in the current directory. The file name must be the same as the function name, so in this case it must be stored as an M-file named **f**, which is the default suggestion when the save icon is selected from the menu bar. Finally, the output, **s**, is defined as **x+3**.

From the command window, or inside a **script M-file**, the function **f** is now available. Now type

```
f(4)
```

and the program returns

```
ans =
     7
```

More complicated functions can be written that require more than one input argument. For example, these lines of code define a function called **g**, with two inputs, **x** and **y**:

```
function output=g(x,y)
% This function multiplies x and y together
% Be sure that x and y have the same size.
a = x.*y;
output = a;
```

You can use the comment lines to let users know what kind of input is required and to describe the function. In this example the function returns the values in the matrix named **output**.

You can also create functions that return more than one output variable. Many of the predefined MATLAB function have options to return more than one result.

For example, **max** returns the maximum value in a matrix, but it can also return the element number where the maximum occurs. Examples of these two references are:

```
max_value = max(A);
[max_value, max_location] = max(A);
```

To achieve a similar result in a user-defined function, make the output a matrix of answers, instead of a single variable. For example, consider the following function that computes the volume and the surface area of a cube, given the length of a side:

```
function [volume, surface] = cube(side)
% This function computes the volume and surface area
% of a cube, give the value of one side.
volume = side.^3;
surface = 6*side.^2;
```

Once saved as **cube** in the current directory, you can use the function to find values of the volume and surface area with statements such as the following:

```
[vol, area] = cube(5)
vol =
     125
area =
      150
```

If you call the **cube** function without specifying both outputs, only the first output will be returned:

```
cube(1.5)
ans =
     3.3750
```

3.6.2 Local Variables

The variables used within function M-files are known as **local variables**. The only way that a function can communicate with the workspace is through input arguments and the output returned. Any variables defined within the function only exist for the function to use. For example, consider the **g** function previously described:

```
function output=g(x,y)
% This function multiplies x and y together
% Be sure that x and y are the same size matrices
a = x .*y;
output = a;
```

The variable `a` is a local variable. It can be used for additional calculations inside the function `g`, but it is not stored in the workspace. To confirm this, clear the workspace and the command window, then call the `g` function:

```
clear, clc
g(10,20)
```

returns

```
g(10,20)
ans =
     200
```

Notice that the only variable stored in the workspace window is **ans**. Not only is **a** not there, but neither is **output**, which is also a local variable.

Just as calculations performed in the command window, or from a script M-file, cannot access variables defined in functions, functions cannot access the variables defined in the workspace. That means that functions must be completely self-contained. The only way they can get information from your program is through the input arguments, and the only way they can deliver information is through the function output.

3.6.3 Naming User-Defined Functions

A function M-file must have the same file name as its function name defined in the first line. For example,

```
function results = velocity(t)
```

might be the first line of a user-defined function. The function's name is **velocity**, and it must be stored in the current directory as **velocity**. Function names need to conform to the same naming conventions as variable names: They must start with a letter; they may only contain letters, numbers and the underscore; and they must not be reserved names. It is possible to give a function the same name as a predefined MATLAB function, in which case the user-defined function will become the default until it is removed from the current directory or the current directory is changed. The MATLAB predefined function is not overwritten; MATLAB just looks in the current directory first for function definitions before it looks into the predefined function files. In general, it is not a good idea to use the same name for a user-defined function as for an existing MATLAB function.

3.6.4 Rules for Writing and Using User-Defined Functions

Writing and using an M-file function require the user to follow very specific rules. These rules are summarized as follows:

- The function must begin with a line containing the word **function**, which is followed by the output argument, an equals sign, and the name of the function. The input arguments to the function follow the name of the function and are enclosed in parentheses. This line distinguishes the function file from a script M-file:

  ```
  function output_name = function_name(input)
  ```

- The first few lines of the function should be comments because they will be displayed if help is requested for the function name:

  ```
  % Add comments to your function for users
  ```

- The only information returned from the function is contained in the output arguments, which are, of course, matrices. Always check to be sure that the function includes a statement that assigns a value to the output argument.
- A function that has multiple input arguments must list the arguments in the function statement, as shown in the following example, which has two input arguments:

  ```
  function error = mse(w,d)
  ```

- A function that is going to return more than one value should show all values to be returned as a vector in the function statement, as in

  ```
  function [volume, surface] = cube(side)
  ```

All output values need to be computed within the function.

- The same matrix names can be used in both a function and the program that references it. No confusion occurs as to which matrix is referenced, because the function and the program are completely separate. However, any values computed in the function, other than the output arguments, are not accessible from the program.
- The special functions **nargin** and **nargout** can be used to determine the number of input arguments and the number of output arguments for a function. Both require a string containing the function name as input. For the **cube** function described earlier,

```
nargin('cube')
```

returns

```
ans =
     1
```

and

```
nargout('cube')
```

returns

```
ans =
     2
```

EXAMPLE 3.6

A FUNCTION TO CONVERT DEGREES CELSIUS TO DEGREES FAHRENHEIT

When working with temperatures, some computations will be easier to perform using degrees Celsius, and other computations will be easier to perform using degrees Fahrenheit. Write a function that will convert degrees Celsius to degrees Fahrenheit.

SOLUTION

1. Problem Statement

Create and test a function called **celsius** to convert degrees Celsius to degrees Fahrenheit.

2. Input/Output Description

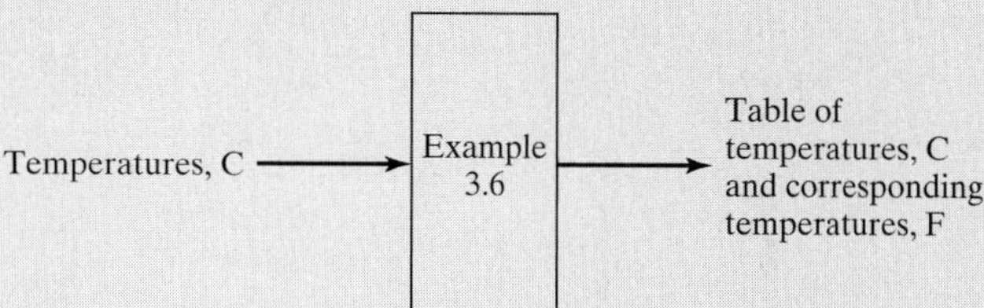

(*continued*)

3. Hand Example

The equation for converting temperatures in Celsius to Fahrenheit is:

$$T_f = 9/5\ T_c + 32$$

If we choose $T_c = 0°C$, then $T_f = 32°F$.

4. MATLAB Solution

```
%--------------------------------------------------------------
% Example 3_6 This program converts temperatures from
% Celsius to Fahrenheit using a user-defined function.
%
clear, clc
%
% Define a vector of temperatures Celsius.
Tc = 0:10:200;
%
% Use a function to convert temperature to Fahrenheit.
Tf = Fahrenheit(Tc);
%
% Generate an output table
Temp(1,:) = Tc';
Temp(2,:) = Tf';
disp('A Conversion Table')
disp('degrees C degrees F')
fprintf('%5.0f %9.2f\n', Temp)
%--------------------------------------------------------------
```

The function called by the program is shown below:

```
%--------------------------------------------------------------
function deg_f = Fahrenheit(Tc)
% This function converts Celsius to Fahrenheit.
%
deg_f = 9/5.*Tc + 32;
%--------------------------------------------------------------
```

Remember that, in order for the script M-file to find the function, it must be in the current directory and must be named **`Fahrenheit.m.`** The program generates the following results in the command window:

```
A Conversion Table
degrees C  degrees F
   0         32.00
  10         50.00
  20         68.00
   :
 180        356.00
 190        374.00
 200        392.00
```

5. Testing

Compare the MATLAB solution to the hand solution. Since the output is a table, it is easy to see that the conversions generated by MATLAB correspond to those calculated by hand. Notice that the titles and column headings were generated with the **disp** function.

SUMMARY

In this chapter, we explored the various predefined MATLAB functions. These functions included general mathematical functions, trigonometric functions, data analysis functions, and random number generation. These functions give us the power to perform operations through a simple function reference. We also learned to write our own functions. This allows us to use operations that we want to repeat without having to repeat the code each time we want to use the operation. It also allows us to keep our MATLAB program shorter, and that is important in maintaining readability.

MATLAB Summary

The following MATLAB summary lists and briefly describes the functions presented in this chapter:

Functions	
`abs`	computes the absolute value
`asin`	computes the inverse sine (arcsine)
`ceil`	rounds to the nearest integer toward positive infinity
`cos`	computes the cosine
`cumprod`	computes a cumulative product of the values in an array
`cumsum`	computes a cumulative sum of the values in an array
`exp`	computes the value of e^x
`fix`	rounds to the nearest integer toward zero
`floor`	rounds to the nearest integer toward minus infinity
`help`	provides documentation for a specified function
`length`	determines the largest dimension of an array
`log`	computes the natural log
`log10`	computes the log base 10
`log2`	computes the log base 2
`max`	finds the maximum value in an array, and determines which element stores the maximum value
`mean`	computes the average of the elements in an array
`median`	finds the median of the elements in an array

Functions	
`min`	finds the minimum value in an array, and determines which element stores the minimum value
`prod`	multiplies the values in an array
`rand`	generates evenly distributed random numbers
`randn`	generates normally distributed (Gaussian) random numbers
`rem`	calculates the remainder in a division problem
`round`	rounds to the nearest integer
`sign`	determines the sign (positive or negative)
`sin`	computes the sine
`sinh`	computes the hyperbolic sine
`size`	determines the number of rows and columns in an array
`sort`	sorts the elements of a vector into ascending order
`sqrt`	calculates the square root of a number
`std`	determines the standard deviation
`sum`	sums the values in an array
`tan`	computes the tangent

KEY TERMS

argument
built-in functions
composition of functions
computer simulation
function
Gaussian random numbers
import data
local variable
mean
Monte Carlo simulation
natural logarithm
nested functions
normal random numbers
random number
seed
standard deviation
uniform random number
user-defined function
median

PROBLEMS

1. Sometimes it is convenient to have a table of sine, cosine, and tangent values instead of using a calculator. Create a table of all three of these trigonometric functions for angles from 0 to 2π in increments of 0.1 radians. Your table should contain a column for the angle, followed by the three trigonometric function values.
2. The range of an object shot at an angle θ with respect to the x axis and an initial velocity v_0 is given by

$$R(\theta) = \frac{v^2}{g}\sin(2\theta) \text{ for } 0 \le \theta \le \frac{\pi}{2} \text{ and neglecting air resistance.}$$

Use $g = 9.9 \text{ m/s}^2$ and an initial velocity of 100 m/s. Show that the maximum range is obtained at $\theta = \pi/4$ by computing the range in increments of 0.05 from $0 \leq \theta \leq \pi/2$. Because you are using discrete angles, you will only be able to determine θ to within 0.05 radians. Remember, **`max`** can be used to return not only the maximum value in an array, but also the element number where the maximum value is stored.

3. MATLAB contains functions to calculate the natural log (`log`), the log base 10 (`log10`) and the log base 2 (`log2`). However, if you want to find a logarithm to another base, for example base *b*, you will have to do the math yourself:

$$\log_b(x) = \frac{\log_e(x)}{\log_e(b)}.$$

What is the log of 10 to the base *b*, when *b* is defined from 2 to 10 in increments of 1?

4. Populations tend to expand exponentially:

$$P = P_0 e^{rt}$$

where P is the current population,

P_0 is the original population,
r is the rate, expressed as a fraction, and
t is the time.

If you originally have 100 rabbits that breed at a rate of 90 percent (0.9) per year, find how many rabbits you will have at the end of 10 years.

5. Chemical reaction rates are proportional to a rate constant, *k*, which changes with temperature according to the Arrhenius equation

$$k = k_0 e^{-Q/RT}$$

For a certain reaction

$$Q = 8{,}000 \text{ cal/mole}$$
$$R = 1.987 \text{ cal/mole K}$$
$$k_0 = 1200 \text{ min}^{-1}$$

find the values of *k* for temperatures from 100 K to 500 K, in 50-degree increments. Create a table of your results.

6. The vector **G** represents the distribution of final grades in a statics course. Compute the mean, median, and standard deviation of *G*. Which better represents the "most typical grade," the mean or the median? Why?

```
G = [68,83,70,75,82,57,5,76,85,62,71,96,78,76,72,75,83,93]
```

Use MATLAB to determine the number of grades in the array. (Do not just count them.)

7. Generate 10,000 Gaussian random numbers with a mean of 80 and standard deviation of 23.5. Use the **`mean`** function to confirm that your array actually has a mean of 80. Use the **`std`** function to confirm that your standard deviation is actually 23.5.

ROCKET ANALYSIS

A small rocket is being designed to make wind shear measurements in the vicinity of thunderstorms. Before testing begins, the designers are developing a simulation of the rocket's trajectory. They have derived the following equation, which they believe will predict the performance of the test rocket, where t is the elapsed time, in seconds:

$$\text{height} = 2.13t^2 - 0.0013t^4 + 0.000034t^{4.751}$$

8. Compute and print a table of time versus height, at 2-second intervals, up through 100 seconds. (The equation will actually predict negative heights. Obviously, the equation is no longer applicable once the rocket hits the ground. For now, do not worry about this physical impossibility; just do the math.)

9. Use MATLAB to find the maximum height achieved by the rocket.

10. Use MATLAB to find the time the maximum height is achieved.

SENSOR DATA

Suppose that a file named **`sensor.dat`** contains information collected from a set of sensors. Each row contains a set of sensor readings, with the first row containing values collected at 0 seconds, the second row containing values collected at 1.0 seconds, and so on.

11. Write a program to read the data file and print the number of sensors and the number of seconds of data contained in the file. (Hint: use the **size** function.)

12. Find both the maximum value and minimum value recorded on each sensor. Use MATLAB to determine at what times they occurred.

13. Find the mean and standard deviation for each sensor, and for all the data values collected.

TEMPERATURE DATA

Suppose you are designing a container to ship sensitive medical materials between hospitals. The container needs to keep the contents within a specified temperature range. You have created a model predicting how the container responds to exterior temperature, and now need to run a simulation.

14. Create a normal distribution of temperatures (Gaussian distribution) with a mean of 70°F, and a standard deviation of 2 degrees, corresponding to 2 hours duration. You will need a temperature for each minute from 0 to 120 minutes.

15. Plot the data on an x-y plot. (Chapter 4 covers labels, so do not worry about them. Recall that the MATLAB function for plotting is **`plot(x,y)`**.)

16. Find the maximum temperature and the minimum temperature.

CHAPTER

4 Plotting

Objectives

After reading this chapter, you should be able to

- create and label two-dimensional plots,
- adjust the appearance of your plots,
- create three-dimensional plots, and
- use the interactive MATLAB plotting tools.

ENGINEERING ACHIEVEMENT: OCEAN DYNAMICS

Waves are generated by wind, earthquakes, storms, and the tides from the gravitational pull of the Sun and the Moon. The energy being transmitted through the water causes the water particles to oscillate in place, but the water itself does not travel from one location to another. This vibration or oscillation can be back and forth in the direction of the energy flow, or it can be in circular orbits along interfaces between water layers with different densities. Waves have crests (high points) and troughs (low points). The vertical distance between a crest and trough is the wave height, and the horizontal distance from crest to crest is the wavelength. Deepwater waves occur where the water depth is greater than one-half the wavelength, and they are often generated by winds at the ocean surface. The water depth does not affect the speed of deepwater waves. Shallow-water waves are those in which the ocean depth is less than 1/20 of the wavelength, and include tide waves. The speed of shallow-water waves is determined by the water depth—the greater the depth, the higher the wave speed. The speed of transitional waves (those in depths between one-half and one-twentieth of the wavelength) are determined by wavelength and water depth. In this chapter, we analyze the interference of waves as different waves come together.

4.1 *X-Y* PLOTS

Large tables of data are difficult to interpret. Engineers use graphing techniques to make the information more accessible. With a graph it is easy to identify trends, pick out highs and lows, and isolate data points that may be measurement or calculation errors. A graph can also be used as a quick check to determine whether or not a computer solution is yielding expected results. The most common plot used by engineers is the ***x-y* plot**. The data that we plot are usually read from a data file, or computed in programs and stored in vectors that we will call x and y. Generally, the x values represent the **independent variable** and the y values represent the **dependent variable**. The y values can be computed as a function of x or the x and y values might be measured in an experiment. In this chapter, we will also present other types of graphs, including bar charts, pie charts, polar plots, histograms, and 3D plots.

4.1.1 Basic *x-y* Plots

Once the **x** and **y** vectors have been defined, MATLAB makes it easy to create plots. The data shown in Table 4.1 were collected from an experiment with a remotely controlled model car. The experiment was repeated 10 times, and we have measured the distance that the car traveled for each trial.

Assume that the trial numbers are stored in a vector called **x** and the distance values are stored in a vector called **y**:

```
x = [1:10];
y = [58.5,63.8,64.2,67.3,71.5,88.3,90.1,90.6,89.5,90.4];
```

To plot these points, we use the **plot** command, with **x** and **y** as arguments:

```
plot(x,y)
```

A graphics window automatically opens, which MATLAB calls Figure 1. The resulting plot is shown in Figure 4.1. (Slight variations in the scaling of the plot may occur, depending on the computer type and the size of the graphics window. We have also specified a thicker line width in the plots in this text to make the plots easier to read.)

Table 4.1 Experimental Distances from 10 Trials

Trial	Distance (ft)
1	58.5
2	63.8
3	64.2
4	67.3
5	71.5
6	88.3
7	90.1
8	90.6
9	89.5
10	90.4

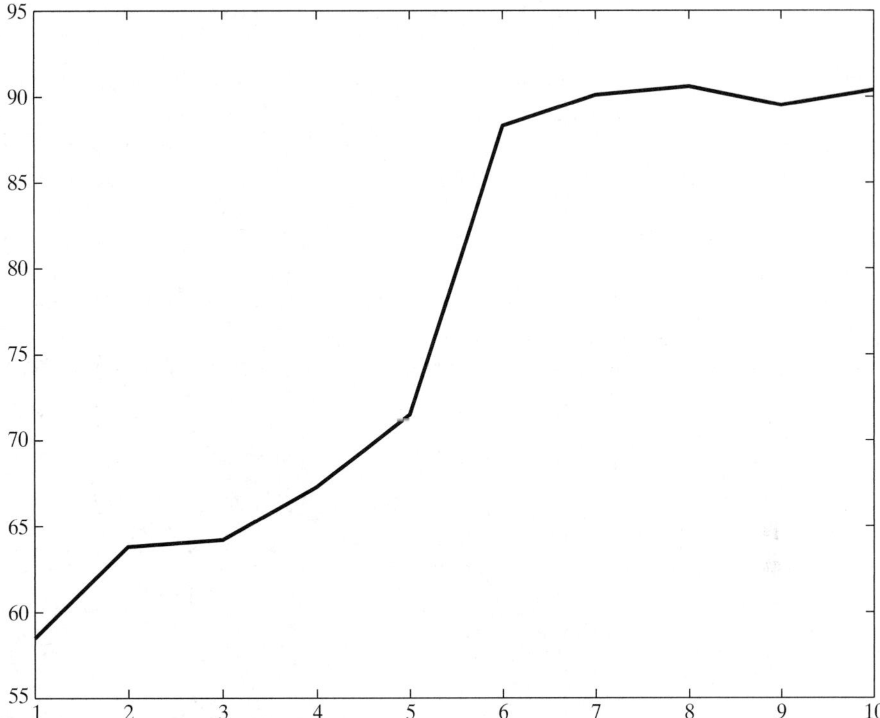

Figure 4.1
Simple plot of distances for 10 trials with a remotely controlled model car.

Good engineering practice requires that we include units and a title in our plot. The following commands add a title, *x*- and *y*-axis labels, and a background grid:

```
plot(x,y),title('Laboratory Experiment 1'),
   xlabel('Trial'),ylabel('Distance, ft'),grid
```

These commands generated the plot in Figure 4.2. As you type these commands into MATLAB, notice that the type color changes to red when you enter a single quote (`'`). This alerts you that you are starting a string. The color changes to purple when you type the final single quote (`'`), indicating that you have completed the string. Paying attention to these visual aids will help you to avoid coding mistakes.

If you are working in the command window, the graphics window may open behind the other windows. Click in the graphics window to display it on top of the command window as shown in Figure 4.3. To continue working, either click in the command window or minimize the graphics window. You can also resize the graphics window to whatever size is convenient for you.

If you request a plot from an M-file, and then continue on with more computations, MATLAB will generate and display the graphics window and then return immediately to execute the rest of the commands in the program. (See Figure 4.3.) If you request a second plot, the graph you created will be overwritten. There are two possible solutions to this problem. You can use the **`pause`** command to temporarily halt the execution of your M-file program, or you can create a second figure using the **`figure`** function.

The **`pause`** command stops the program execution until any key is pressed. If you want to pause for a specified number of seconds, use the **`pause(n)`** command, which will cause an execution pause for **`n`** seconds before continuing.

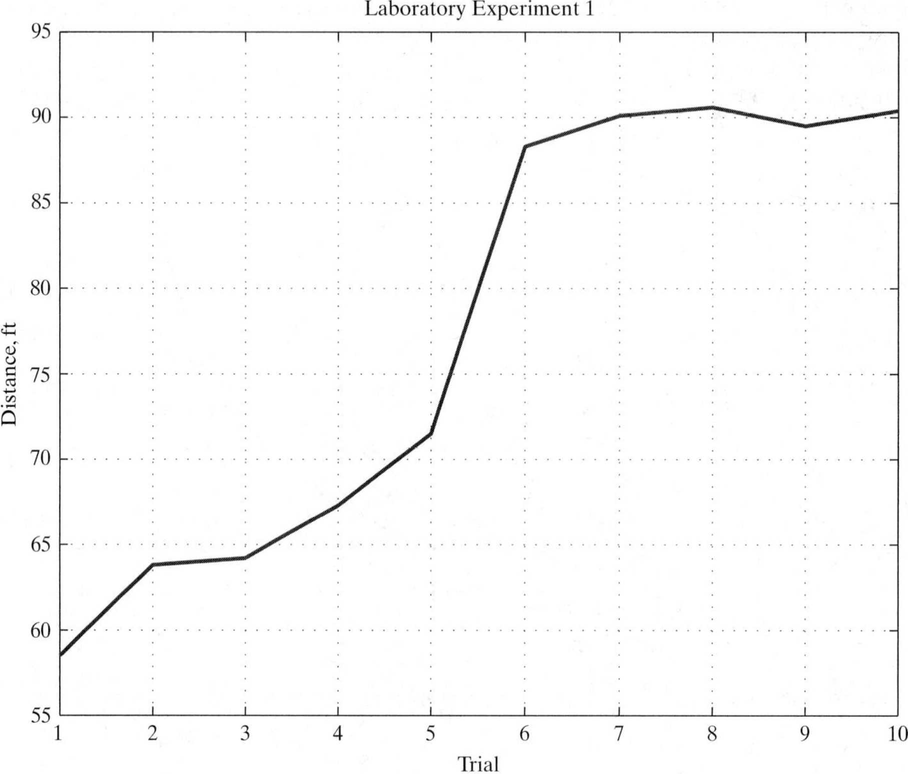

Figure 4.2
Simple plot with a grid, title, and labels.

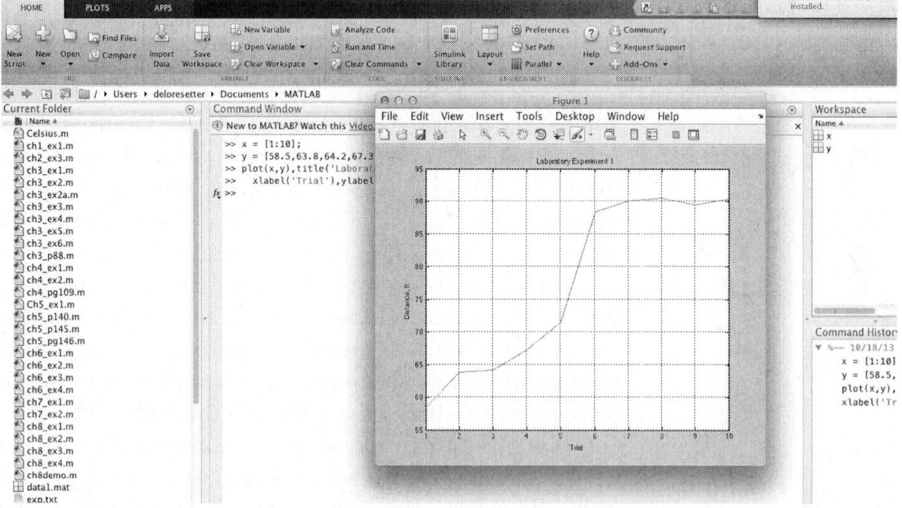

Figure 4.3
The graphics window on top of the command window. (This graphics window can be resized to a convenient size.)

The **`figure`** command allows you to open a new figure window. The next time you request a plot, it will be displayed in this new window. For example,

```
figure(2)
```

opens a window named Figure 2, which then becomes the window used for subsequent plotting. Other figure windows will be behind the current figure window.

Creating a graph with more than one set of data plotted in it can be accomplished in several ways. By default, the execution of a second **`plot`** statement will erase the first plot. However, you can layer plots on top of one another by using the **`hold on`** command. Execute the following statements to create a plot with both functions plotted on the same graph, as shown in Figure 4.4:

```
x = 0:pi/100:2*pi;
y1 = cos(x*4);
plot(x,y1),title('Two Plots on the Same Axis'),
   xlabel('angle, radians'),grid
y2 = sin(x);
hold on;
plot(x,y2)
```

Semicolons are optional on both the plot statements and the **`hold on`** statement. MATLAB will continue to layer the plots until the **`hold off`** command is executed:

```
hold off
```

Another way to create a graph with multiple lines is to request both lines in a single **`plot`** command. To illustrate, consider this MATLAB command:

```
plot(X,Y,W,Z)
```

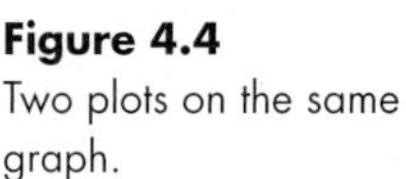
Figure 4.4
Two plots on the same graph.

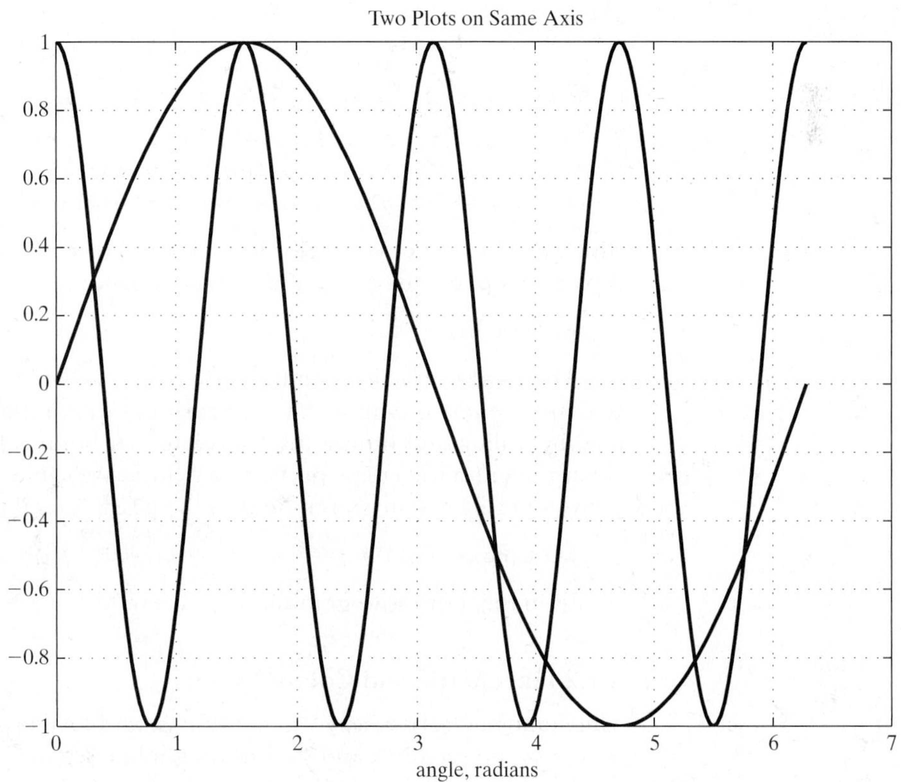

In this statement, **X** and **Y** represent one set of values to be plotted, and **W** and **Z** form a second set of values. Using the data from the previous examples,

```
plot(x,y1,x,y2)
```

produces the same graph as Figure 4.4, with one exception—the two lines are different colors. MATLAB uses a default plotting color (blue) for the first line drawn with a **plot** command. When using the **hold on** approach, each line is drawn in a separate **plot** command, and thus is the same color. By requesting two lines in a single command, the second line defaults to green, allowing the user to distinguish between the two plots.

If the **plot** command is used with a single matrix argument, MATLAB draws a separate line for each column of the matrix. The *x*-axis is labeled with the row index vector 1:*k*, where *k* is the number of rows in the matrix. This produces an evenly spaced plot, sometimes called a **line plot**. If the function is called with two arguments, one a vector and the other a matrix, MATLAB successively plots a line for each row in the matrix. For example, combine **y1** and **y2** into a single matrix **y3**, and plot **y3** versus **x**:

```
y3 = [y1;y2];
plot(x,y3)
```

This creates the same plot as Figure 4.4, with each line a different color.

Here is another more complicated example:

```
x = 0:pi/100:2*pi;
y1 = cos(x)*2;
y2 = cos(x)*3;
y3 = cos(x)*4;
y4 = cos(x)*5;
z = [y1; y2; y3; y4];
plot(x,y1,x,y2,x,y3,x,y4),
    title('Multiple Cosine Functions'),
    xlabel('angle, radians'),ylabel('amplitude'), grid
```

which produces the result shown in Figure 4.5. The same result is obtained if you replace the **plot** command with this command:

```
plot(x,z)
```

The **peaks** function is a function of two variables that produces sample data that are useful for demonstrating certain graphing functions. (The data are created by scaling and translating Gaussian distributions.) Calling **peaks** with a single argument **n** will create an $n \times n$ matrix. We can use peaks to demonstrate the power of using a matrix argument in the **plot** function. Hence,

```
plot(peaks(100))
```

results in the impressive graph in Figure 4.6.

4.1.2 Line, Mark, and Color Options

The command **plot(x,y)** generates an *x-y* plot that connects the points represented by the vectors **x** and **y** with straight-line segments. Graphs that appear to be curved are actually created by using a large number of points, but still connecting

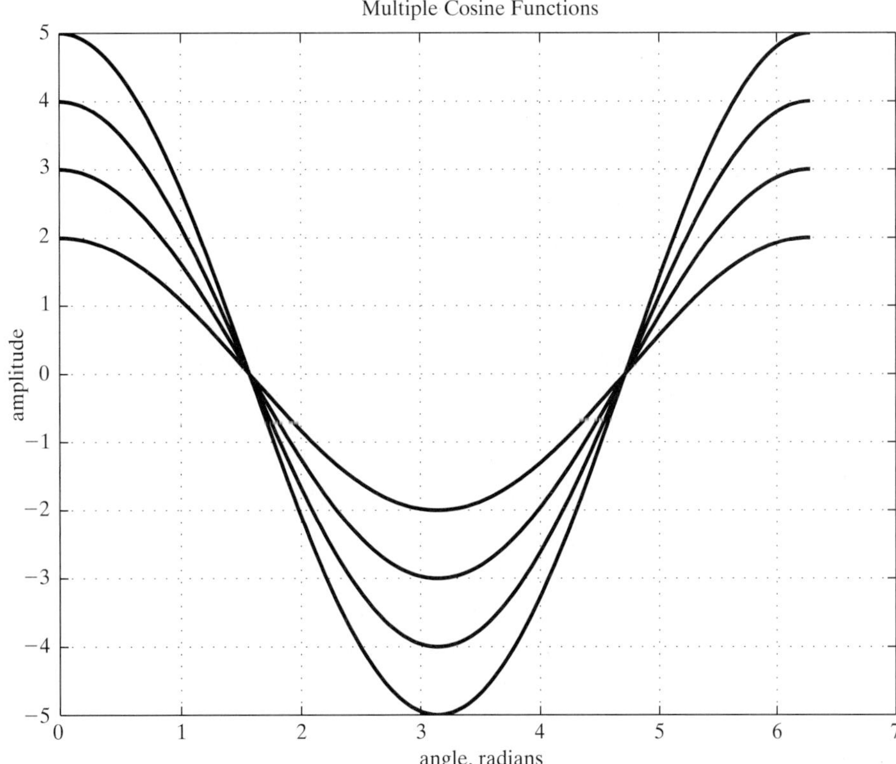

Figure 4.5
Four plots on the same graph.

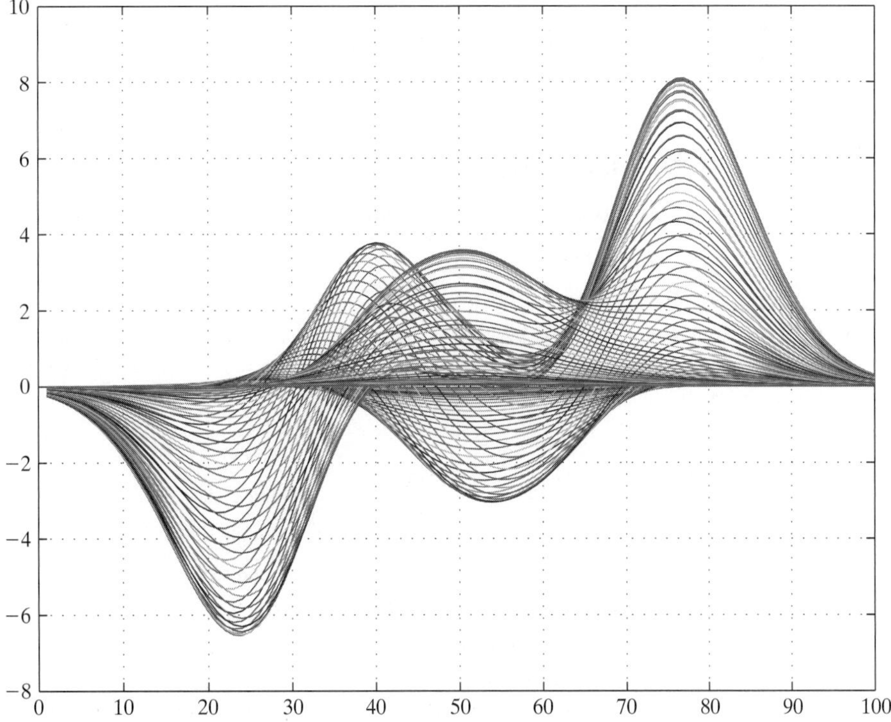

Figure 4.6
The **peaks** function.

them with very short straight-line segments. In the default mode, MATLAB uses a blue line for the first set of data plotted, and changes the color for subsequent sets of data. Although the graph is created by connecting data points, those points are not shown on the graph in the default mode.

You can change the appearance of your plots by selecting user-defined **line styles** and **line colors**, and by choosing to show the data points on the graph with user-specified **mark styles**. By typing

```
help plot
```

in the command window, you can determine the choices available to you. You can select solid (the default), dashed, dotted, and dash-dot line styles, and you can choose to show the points. The choices include plus signs, stars, circles, and x-marks, among others. There are seven different color choices. See Table 4.2 for a complete list.

The following commands illustrate the use of line, color, and mark styles.

```
x = [1:10];
y = [58.5,63.8,64.2,67.3,71.5,88.3,90.1,90.6,89.5,90.4];
plot(x,y,':ok')
```

The resulting plot (see Figure 4.7) consists of a dotted line, data points marked with circles, and drawn in black. The indicators were listed inside a string enclosed in single quotes. The order in which they are entered is arbitrary, and does not affect the output.

To specify line, mark, and color styles for multiple lines, add a string containing the choices after each pair of data. If the string is not included, the defaults are used. For example,

```
plot(x,y,':ok',x,y*2,'--xr',x,y/2,'-b')
```

results in the graph shown in Figure 4.8. (The plots shown in Figures 4.7 and 4.8 also included commands for titles, x-labels, y-labels, and grids.)

Table 4.2 Line, Mark, and Color Options

Line Type	Indicator	Mark	Indicator	Color	Indicator
solid	-	point	.	blue	b
dotted	:	circle	o	green	g
dash-dot	-.	x-mark	x	red	r
dashed	--	plus	+	cyan	c
		star	*	magenta	m
		square	s	yellow	y
		diamond	d	black	k
		triangle down	v		
		triangle up	^		
		triangle left	<		
		triangle right	>		
		pentagram	p		
		hexagram	h		

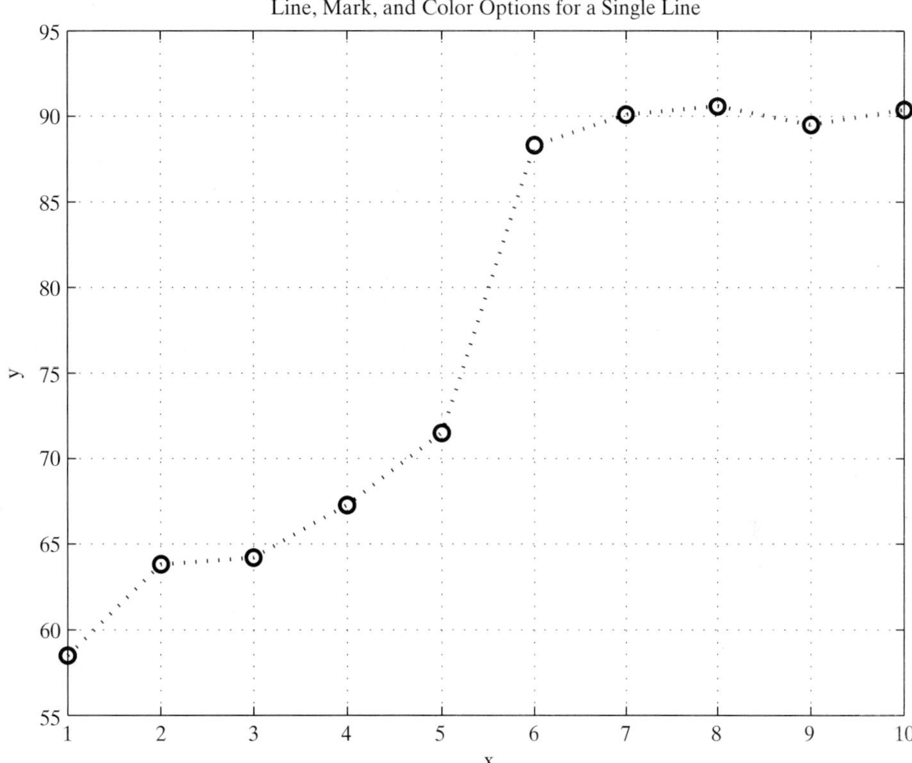

Figure 4.7
Plot with adjusted line, mark, and color type.

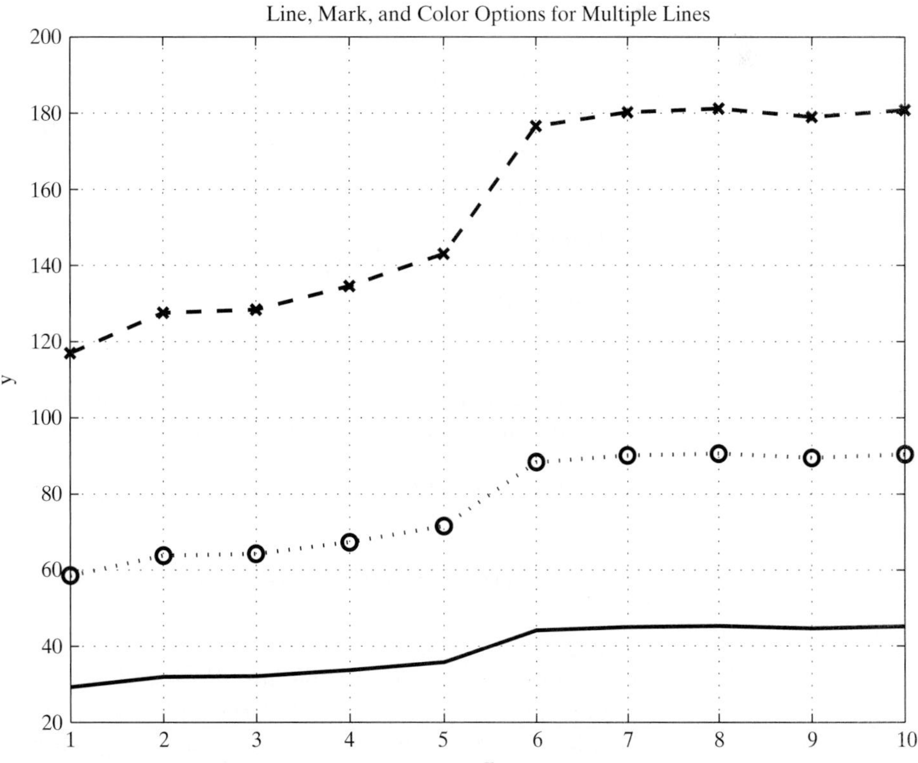

Figure 4.8
Multiple plots with adjusted line, mark, and color type.

4.1.3 Axes Scaling

MATLAB automatically scales the **axes** to fit the data values. However, you can override this scaling with the `axis` command. There are several forms of the `axis` command:

`axis`	Freezes the current `axis` scaling for subsequent plots. A second execution of the command returns the system to automatic scaling.
`axis(v)`	Specifies that the axis to be used is a four-element vector `v` that contains the scaling values `[xmin,xmax,ymin,ymax]`.

4.1.4 Annotating Plots

MATLAB offers several additional functions that allow you to **annotate** your plots. The `legend` function allows you to add a **legend** to your graph. It shows a sample of the line and lists the string(s) you have specified:

```
legend('string1','string 2',...)
```

The `text` function allows you to add a **text box** to the graph. The box is placed at the specified *x* and *y* coordinates, and contains the string value specified:

```
text(x_coordinate,y_coordinate,'string')
```

The following code modifies the graph from Figure 4.8 with both a legend and a text box, in addition to the title, and *x* and *y* labels:

```
plot(x,y,':ok',x,y*2,'--xr',x,y/2,'-b'),
   legend('line 1','line 2','line 3'),
   text(1,100,'Label plots with the text command'),
   xlabel('x'),ylabel('y'),title('Annotated Plot'), grid
```

The results are shown in Figure 4.9.

4.1.5 Other Types of Two-Dimensional Plots

Although simple *x-y* plots are the most common type of engineering plot, there are many other ways to represent data. Depending on the situation, these techniques may be more appropriate than an *x-y* plot.

Polar Plots

MATLAB allows you to generate a **polar plot** using polar coordinates with this command:

```
polar(theta,rho)
```

where `theta` is the angle (in radians) and `rho` is the radial distance. For example, the following statements generate the plot in Figure 4.10:

```
x = 0:pi/100:pi;
y = sin(x);
polar(x,y),
   title('Sine Function in Polar Coordinates')
```

Logarithmic Plots

For most plots that we generate, the *x*- and *y*-axes are divided into equally spaced intervals. These plots are called **linear plots**. Occasionally, we may want to use a logarithmic scale on one or both of the axes. A logarithmic scale (base 10) is convenient

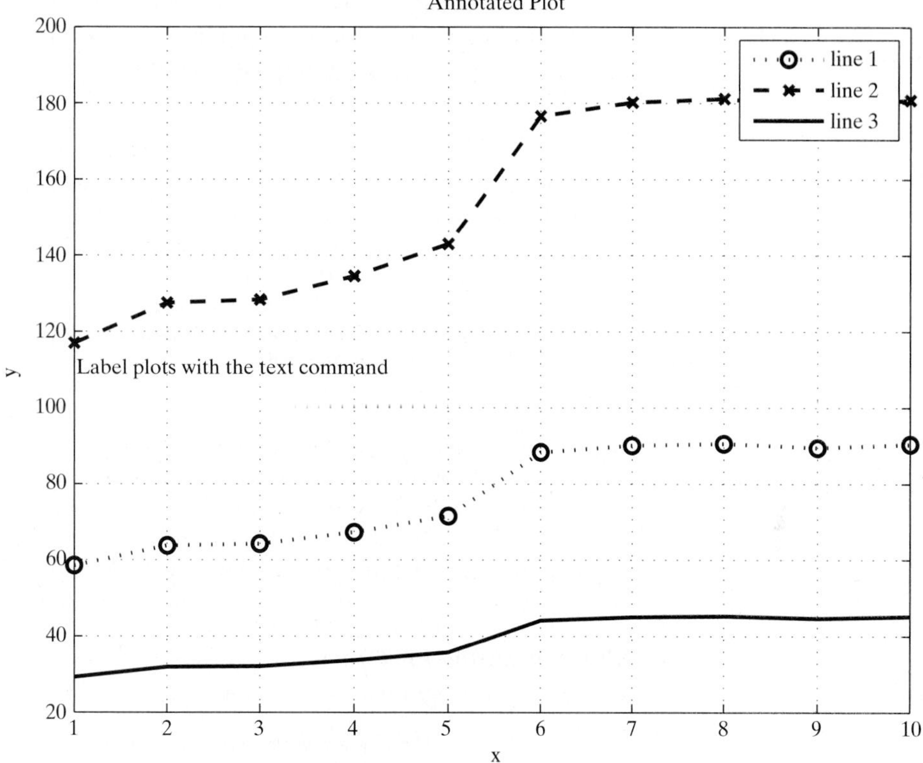

Figure 4.9
Graph annotated with a legend, a text box, title, and x and y labels.

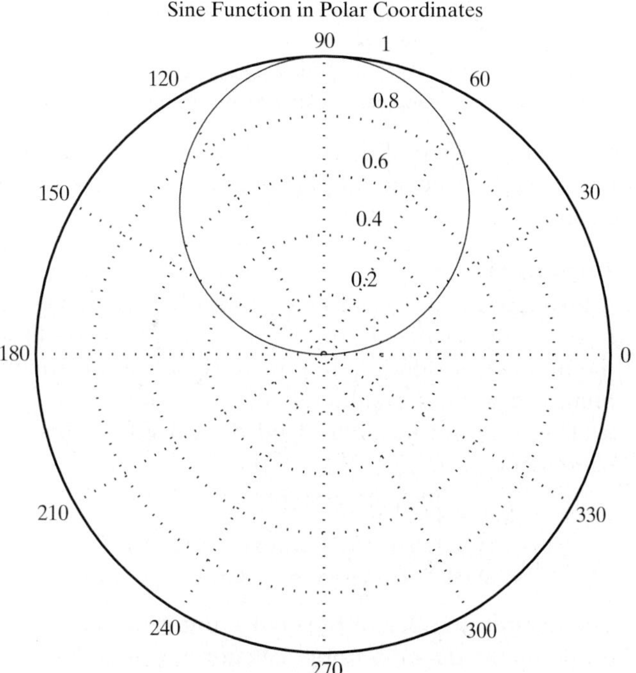

Figure 4.10
A polar plot of the sine function.

when a variable ranges over many orders of magnitude, because the wide range of values can be graphed without compressing the smaller values. **Logarithmic plots** are also useful for representing data that varies exponentially.

The MATLAB commands for generating linear and logarithmic plots of the vectors **x** and **y** are the following:

`plot(x,y)`	Generates a linear plot of the vectors **x** and **y**.
`semilogx(x,y)`	Generates a plot of the values of **x** and **y** using a logarithmic scale for **x** and a linear scale for **y**.
`semilogy(x,y)`	Generates a plot of the values of **x** and **y** using a linear scale for **x** and a logarithmic scale for **y**.
`loglog(x,y)`	Generates a plot of the vectors **x** and **y**, using a logarithmic scale for both **x** and **y**.

It is important to recognize that the logarithms of a negative value and of zero do not exist. Therefore, if the data to be plotted in a semilog plot or a log–log plot contain negative or zero values, a warning message will be printed by MATLAB informing you that these data points have been omitted from the data plotted.

Each of the commands for logarithmic plotting can be executed with one argument, as we saw in `plot(y)` for a linear plot. In these cases, the plots are generated with the values of the indices of the vector **y** used as **x** values.

Bar Graphs and Pie Charts

Bar graphs, **histograms**, and **pie charts** are popular techniques for reporting data:

`bar(x)`	When **x** is a vector, **bar** generates a vertical bar graph. When **x** is a two-dimensional matrix, this function groups the data by row.
`barh(x)`	When **x** is a vector, **barh** generates a horizontal bar graph. When **x** is a two-dimensional matrix, this function groups the data by row.
`bar3(x)`	Generates a three-dimensional bar chart.
`bar3h(x)`	Generates a three-dimensional horizontal bar chart.
`pie(x)`	Generates a pie chart. Each element in the matrix is represented as a slice of the pie.
`pie3(x)`	Generates a three-dimensional pie chart. Each element in the matrix is represented as a slice of the pie.

Examples of some of these graphs were generated for Figure 4.11. In the next section, we will cover the subplot command which was used to display four plots in one figure.

Histograms

A **histogram** is a special type of graph particularly relevant to statistics because it shows the distribution of a set of values. In MATLAB, the histogram computes the number of values falling in 10 bins that are equally spaced between the minimum and maximum values, from the set of values. To illustrate, we define a matrix **x** using random numbers, and then display a histogram of the data with these commands:

```
x = randn(1,10000);
hist(x),title('Histogram with 10 Bins'),
   xlabel('x'),ylabel('Count'), grid
```

The default number of bins is 10, but if we have a large data set we may want to divide up the data into more categories (bins). For example, to create a histogram with 25 bins for the previous example, the command would be

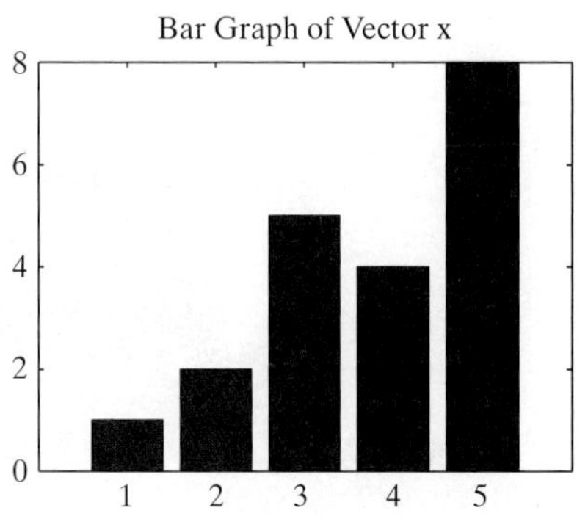

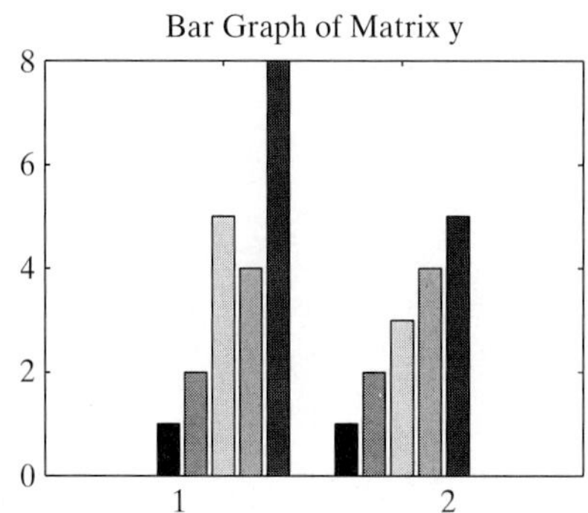

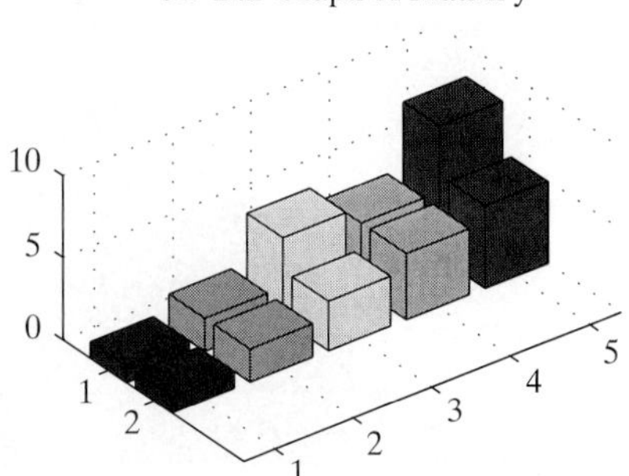

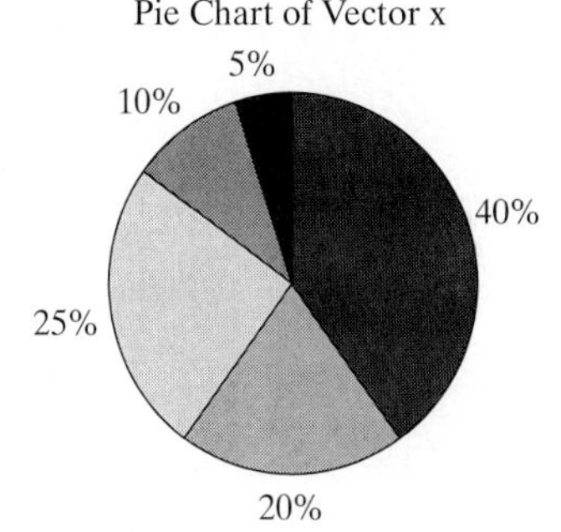

Figure 4.11
Plots of bar graphs and pie charts.

```
hist(x,25),title('Histogram with 25 Bins'),
   xlabel('x'),ylabel('Count'), grid
```

Two typical histograms are displayed in Figure 4.12. The actual histograms depend on the value of the random seed when the random numbers were generated.

4.1.6 Subplots

The **subplot** command allows you to split the graphing window into subwindows. Two subwindows can be arranged either top to bottom or left to right. A four-window split has two subwindows at the top and two subwindows at the bottom. The arguments to the subplot command are three integers, *m*, *n*, and *p*. The digits *m* and *n* specify that the graph window is to be split into an *m*-by-*n* grid of smaller windows, and the digit *p* specifies the *p*th window for the current plot. The windows are numbered from left to right, top to bottom. Therefore, the following commands specify that the graph window is to be split into a top plot and a bottom plot, and the current plot is to be placed in the top subwindow:

```
subplot(2,1,1)
```

The following commands from an M-file generated the plot from Figure 4.11:

```
% Compute values for vector x and matrix y.
x = [1,2,5,4,8];
y = [x; 1:5];
%
% Generate four graphs and display in one plot.
subplot(2,2,1),
   bar(x),title('Bar Graph of Vector x'),
```

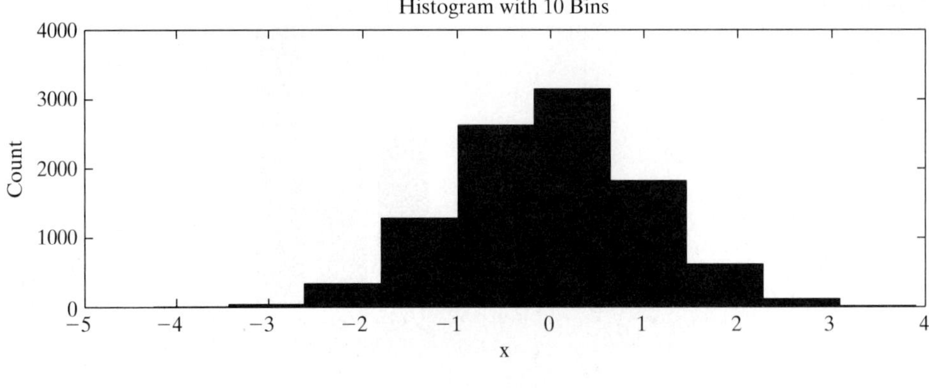

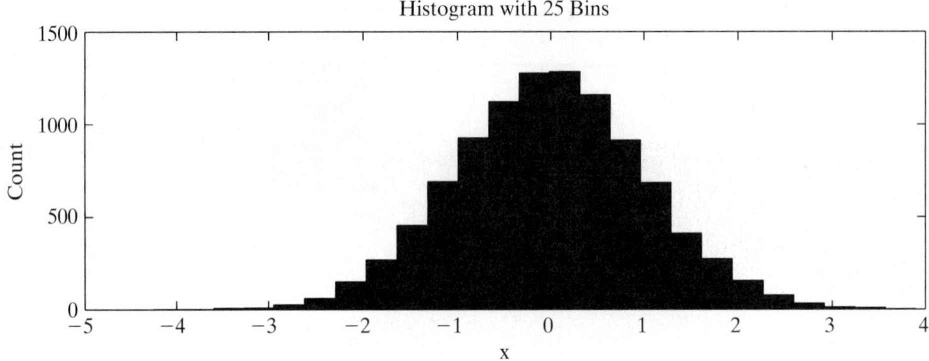

Figure 4.12 Histograms of random data.

```
subplot(2,2,2),
   bar(y),title('Bar Graph of Matrix y'),
subplot(2,2,3), grid,
   bar3(y),title('3D Bar Graph of Matrix y'),
subplot(2,2,4),
   pie(x),title('Pie Chart of Vector x');
```

Figure 4.13 contains four plots that illustrate the **subplot** command, along with the linear and logarithmic plot options. This figure was generated with the following statements from an M-file:

```
% Generate plots of a polynomial
%
x = 0:0.5:50;
y = 5*x.^2;
subplot(2,2,1),plot(x,y),
   title('linear/linear'),
   ylabel('y'),grid,
subplot(2,2,2),semilogx(x,y),
   title('log/linear'),
   ylabel('y'),grid,
subplot(2,2,3),semilogy(x,y),
   title('linear/log'),
   xlabel('x'),ylabel('y'),grid,
subplot(2,2,4),loglog(x,y),
   title('log/log'),
   xlabel('x'),ylabel('y'),grid
```

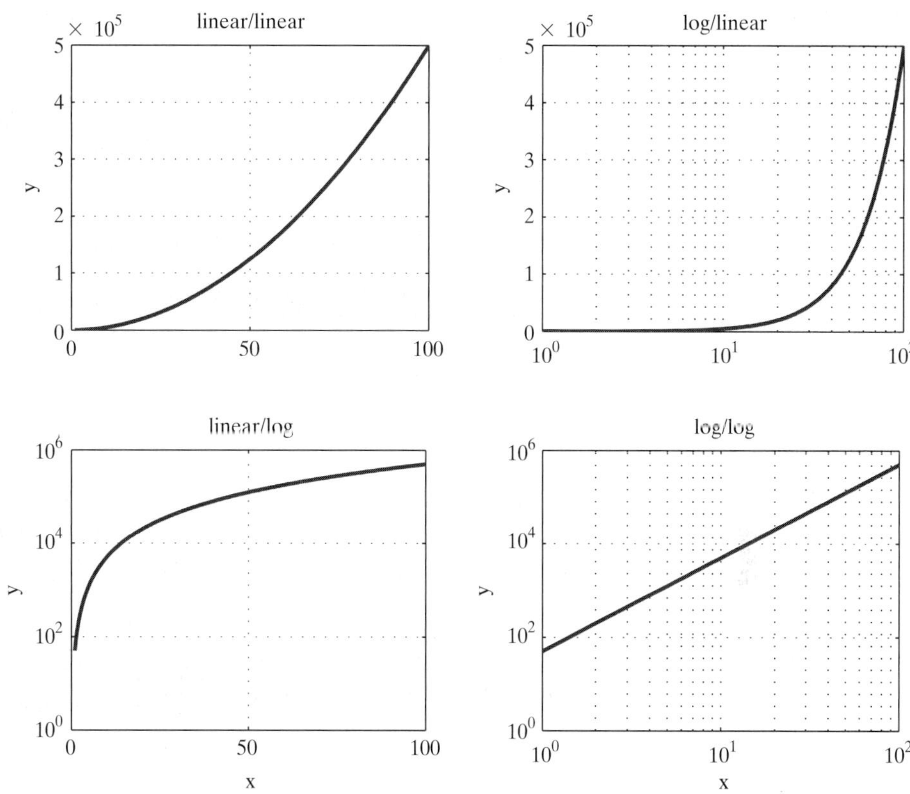

Figure 4.13
Linear and logarithmic plots.

The indenting is intended to make the code easier to read—MATLAB ignores white space. As a matter of style, notice that only the bottom two subplots have *x*-axis labels.

EXAMPLE 4.1

UDF ENGINE PERFORMANCE

Use the following equations to calculate the velocity and acceleration of the advanced turboprop engine called an unducted fan (UDF), and display the results as a table of time, velocity, and acceleration:

$$\text{velocity} = 0.00001\ \text{time}^3 - 0.00488\ \text{time}^2 + 0.75795\ \text{time} + 181.3566$$
$$\text{acceleration} = 3 - 0.000062\ \text{velocity}^2$$

In addition create an *x-y* plot with two lines, one for velocity and one for acceleration.

SOLUTION

1. Problem Statement

Calculate the velocity and acceleration, using a script M-file.

2. Input/Output Description

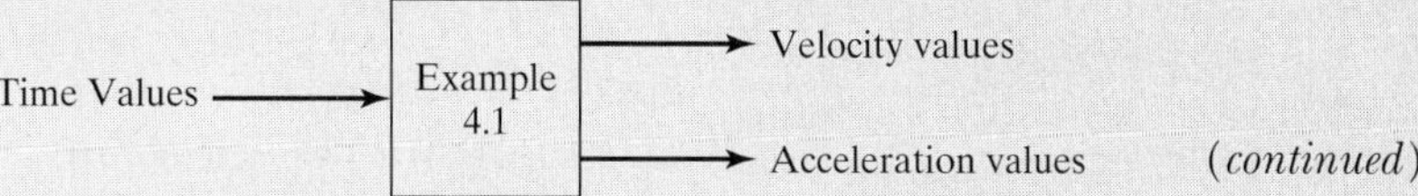

(continued)

3. Hand Example

Solve the equations stated in the problem for time = 10 seconds and 100 seconds:

$$\begin{aligned} \text{velocity} &= 0.00001\ \text{time}^3 - 0.00488\ \text{time}^2 \\ &\quad + 0.75795\ \text{time} + 181.3566 \\ &= 188.46\ \text{m/s (at 10 s)} \\ &= 218.35\ \text{m/s (at 100 s)} \end{aligned}$$

$$\begin{aligned} \text{acceleration} &= 3 - 0.000062\ \text{velocity}^2 \\ &= .798\ \text{m/s}^2\ \text{(at 10 s)} \\ &= 0.04404\ \text{m/s}^2\ \text{(at 100 s)} \end{aligned}$$

4. MATLAB Solution

```
%-----------------------------------------------------------------
% Example 4_1 This program generates velocity and
% acceleration values for a UDF aircraft test.
%
clear, clc
%
% Define the time matrix.
time = 0:10:120;
%
% Calculate the velocity and acceleration matrices.
velocity = 0.00001*time.^3 - 0.00488*time.^2 ...
           + 0.75795*time + 181.3566;
acceleration = 3 - 6.2e-5*velocity.^2;
%
% Display the results in a table
disp('Time, Velocity, Acceleration')
disp('time', velocity', acceleration'])
%
% Create individual x-y plots
figure(1)
subplot(2,1,1),plot(time,velocity),title('UDF Velocity'),grid,
       xlabel('time, s'),ylabel('velocity, m/s')
subplot(2,1,2),plot(time,acceleration),title('UDF Acceleration'),
       xlabel('time, s'),ylabel('acceleration, m/s^2'),grid
%
% Use plotyy to create a scale on each side of a single plot.
figure(2)
plotyy(time,velocity,time,acceleration),grid,
       title('UDF Aircraft Performance'),
       xlabel('time, s'),
       ylabel('velocity, m/s (left); acceleration, m/s^2 (right)')
%-----------------------------------------------------------------
```

The results are returned to the command window:

```
Time, Velocity, Acceleration

        0  181.3566  0.9608
  10.0000  188.4581  0.7980
  20.0000  194.6436  0.6511
  30.0000  199.9731  0.5207
  40.0000  204.5066  0.4070
  50.0000  208.3041  0.3098
  60.0000  211.4256  0.2286
  70.0000  213.9311  0.1625
  80.0000  215.8806  0.1105
  90.0000  217.3341  0.0715
 100.0000  218.3516  0.0440
 110.0000  218.9931  0.0266
 120.0000  219.3186  0.0178
```

Two graphics windows will open. (See Figures 4.14 and 4.15.)

Figure 4.15 is a bit more complicated. We used the **`plotyy`** function instead of the **`plot`** function to force the addition of a second scale on the right-hand side of the plot. We needed this because velocity and acceleration have different units.

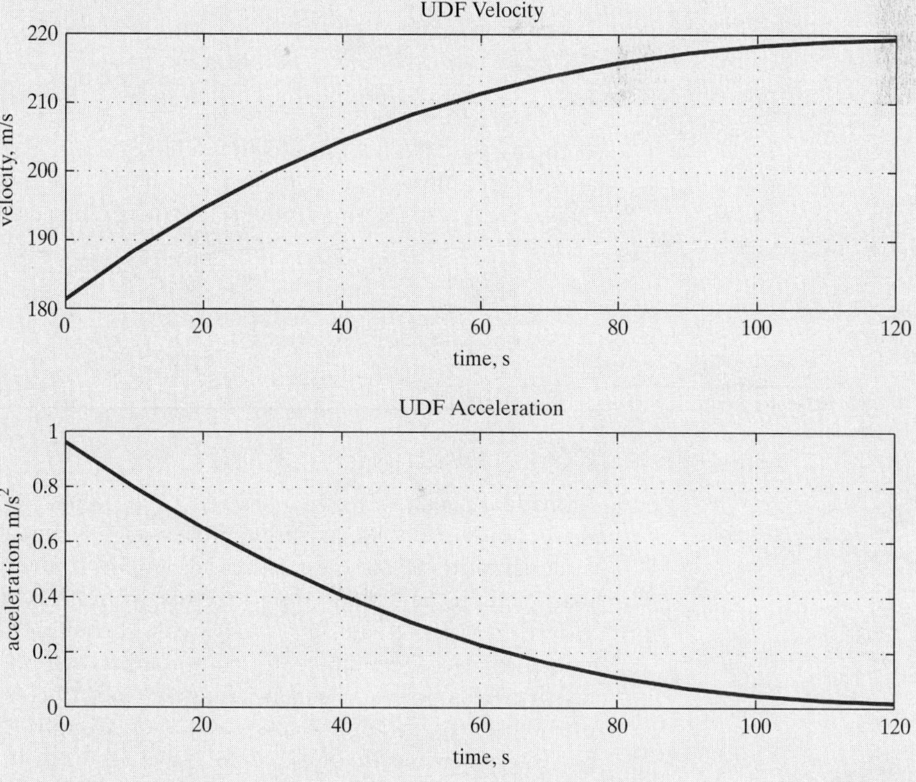

Figure 4.14
Velocity and acceleration of an unducted fan aircraft.

(*continued*)

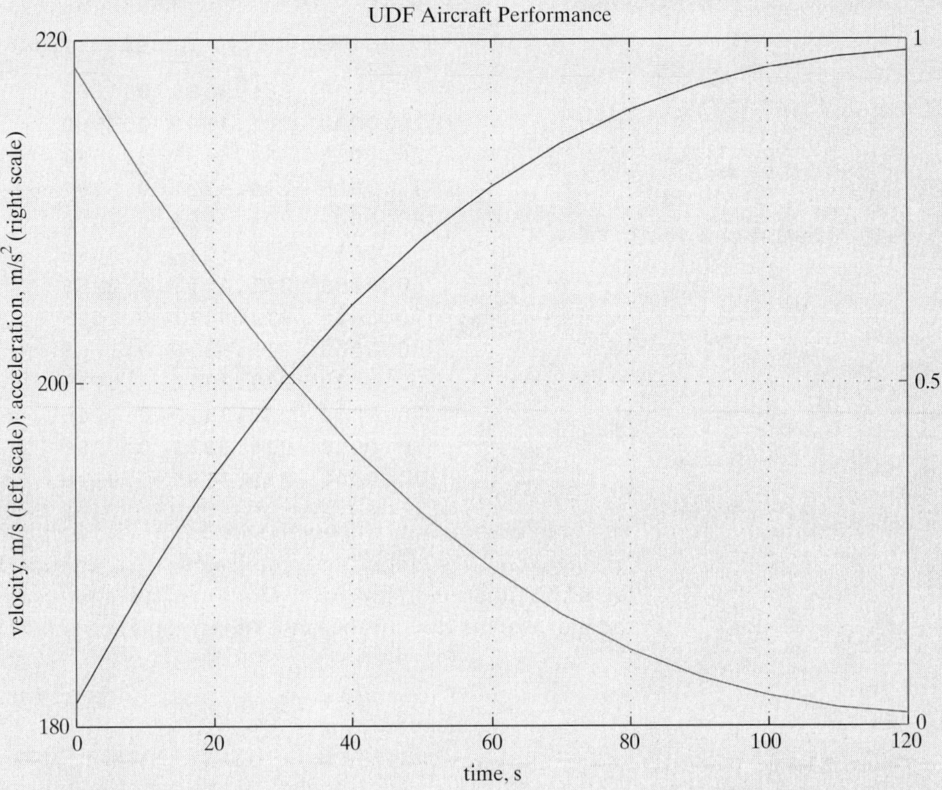

Figure 4.15
Plot using two *y* axes.

5. Testing

Compare the MATLAB results to the hand example results. Notice that the velocity and acceleration calculated from the hand example and the MATLAB solution match. MATLAB automatically scales your graphs using the data plotted.

EXAMPLE 4.2

OCEAN WAVE INTERACTION

Individual ocean waves have crests (high points) and troughs (low points). The vertical distance between a crest and a trough is the wave height, and the horizontal distance from crest-to-crest is the wavelength. In the ocean, waves are generated by many sources, and they come together from many directions. The interaction of individual waves creates a set of waves with varying peaks and troughs. If we consider two waves at a time, the combination can be **constructive interference** in which crests occur at the same time and the troughs occur at the same time, and thus the result has higher crests and lower troughs. The combination is a **destructive interference** if the crest of one wave occurs at the same time as the trough of another wave, because the sum can cancel the highs and lows. In mixed interference, the two waves have different lengths and heights; thus the sum is more complicated because it can now have components of both constructive and destructive interference.

To investigate the interference patterns of two waves, we now develop a program that will allow the user to enter the wave period (in seconds) and wave height from two different waves. The program will then plot the two original waves and their sum. The ocean wave is modeled by a **sinusoid** which is a form of a sine function that is expressed as a function of time instead of a function of angle. A simple sinusoid has the following equation:

$$s(t) = A\sin(2\pi ft + \varphi)$$

where A is the amplitude,
f is the frequency in cycles per second (or hertz), and
φ is the phase shift in radians.

The period of a sinusoid is $1/f$ in seconds.

Consider the following three functions:

$$\begin{aligned} \text{s1}(t) &= 3\sin(2\pi t),\\ \text{s2}(t) &= 5\sin(2\pi(0.2)t),\\ \text{s3}(t) &= 5\sin(2\pi(0.5)t). \end{aligned}$$

Function s1 has an amplitude of 3, and functions s2 and s3 have amplitudes of 5. Function s1 has a frequency of 1 Hz and thus a period of 1 s; functions s2 has a frequency of 0.2 Hz and thus a period of 5 s; and function s3 has a frequency of 0.5 Hz and thus a period of 2 s. The phase shift of all three functions is zero. Figure 4.16 contains plots of these three sinusoids.

Figure 4.16
Plots of three sinusoids.

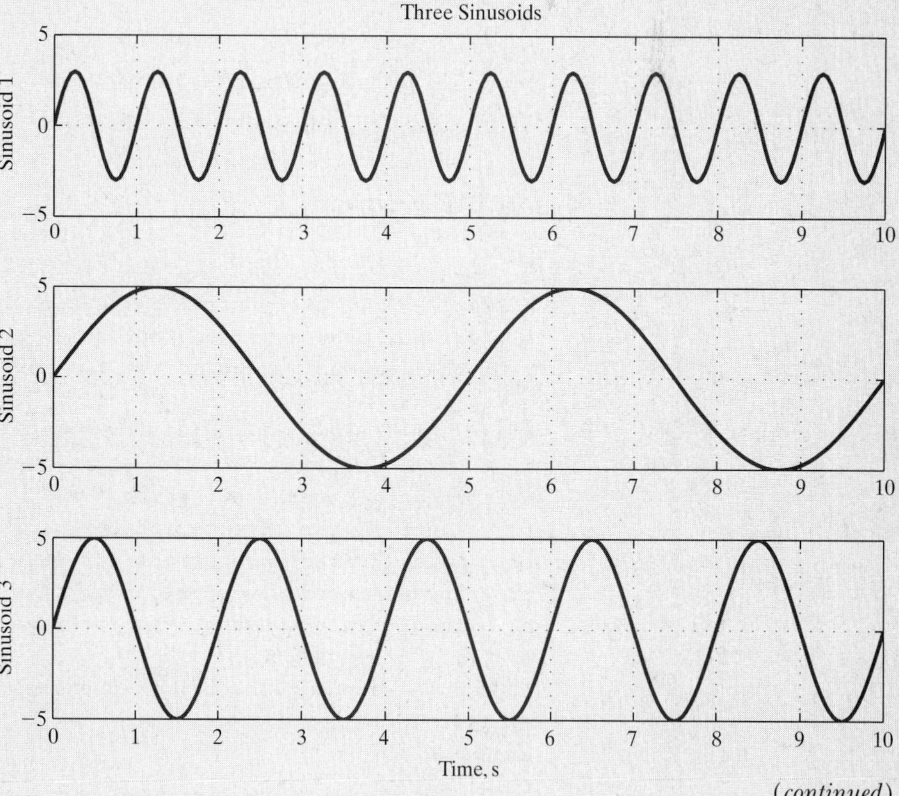

(*continued*)

We now want to write a MATLAB program that will allow the user to enter the period and wave height for two waves. Compute 200 points of the sum of the two waves over a period of time that is four times the maximum of the periods of the two waves.

SOLUTION

1. Problem Statement

Plot the sum of two waves over a period that is four times the maximum period of the two waves.

2. Input/Output Description

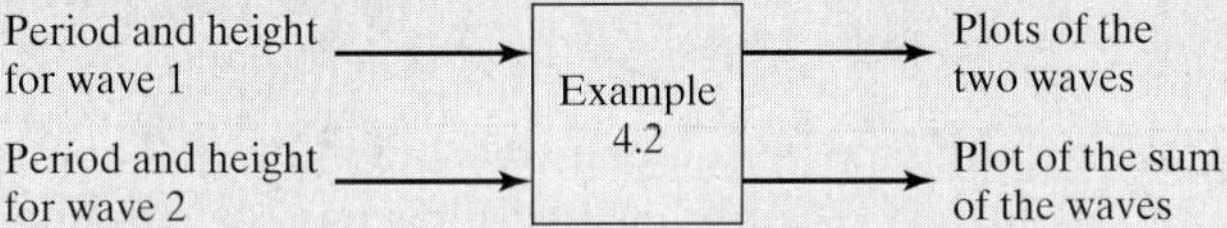

3. Hand Example

Assume that we use s1 and s2 from the previous examples. Thus we have a wave 1 with a period of 1 second and height of 6 ft; wave 2 has a period of 5 seconds and height of 10 ft. The sum of the two waves has the following equation:

$$s(t) = 3\sin(2\pi t) + 5\sin(0.4\pi t)$$

The maximum period is 5 s, so we want to compute values from 0 s to 20 s. The time increment for 200 points is then 20/200, or 0.1 s. The first three values of the sum are then the following:

$$s(0) = 3\sin(0) + 5\sin(0) = 0$$

$$s(0.1) = 3\sin(0.2\pi) + 5\sin(0.04\pi) = 1.76 + 0.63 = 2.39$$

$$s(0.2) = 3\sin(0.4\pi) + 5\sin(0.08\pi) = 2.85 + 1.24 = 3.54$$

4. MATLAB Program

```
%-----------------------------------------------------------
% Example 4_2 This program plots two waves and
% their sum over a period four times the length
% of the maximum period of the two waves.
%
clear, clc
%
% Input data for the two waves.
t1 = input('Enter wave period for wave 1: ');
h1 = input('Enter wave height for wave 1: ');
t2 = input('Enter wave period for wave 2: ');
h2 = input('Enter wave height for wave 2: ');
%
% Compute wave amplitudes, frequencies, and combined period.
amp1 = h1/2;
amp2 = h2/2;
freq1 = 1/t1;
```

```
freq2 = 1/t2;
period = 4*max(t1,t2);
time_incr = period/200;
%
% Compute wave values.
k = 0:199;
t = k*time_incr;
w1 = amp1*sin(2*pi*freq1*t);
w2 = amp2*sin(2*pi*freq2*t);
w3 = w1 + w2;
%
% Plot the two waves and their sum.
subplot(3,1,1),plot(t,w1),grid,
   title('Two Waves and Their Sum'),ylabel('Wave 1, ft'),
subplot(3,1,2),plot(t,w2),grid,ylabel('Wave 2, ft'),
subplot(3,1,3),plot(t,w3),grid,ylabel('Wave Sum, ft'),
   xlabel('Time, s')
%-----------------------------------------------------------------
```

5. Testing

We first test the program using the data from the hand example. The plot for this data is shown in Figure 4.17, and matches the values computed in our hand example. Figure 4.18 contains the output using the first and third sinusoids, and Figure 4.19 contains the output using the second and third sinusoids.

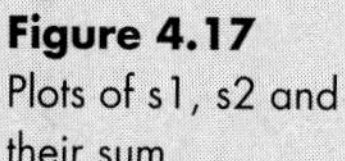
Figure 4.17
Plots of s1, s2 and their sum.

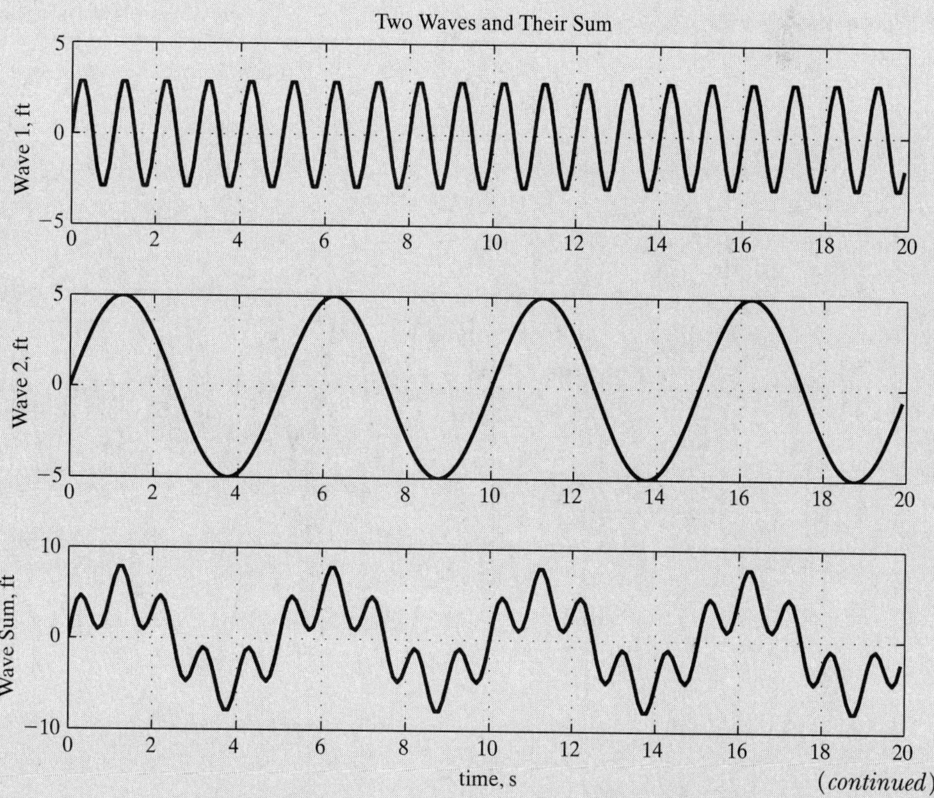

(*continued*)

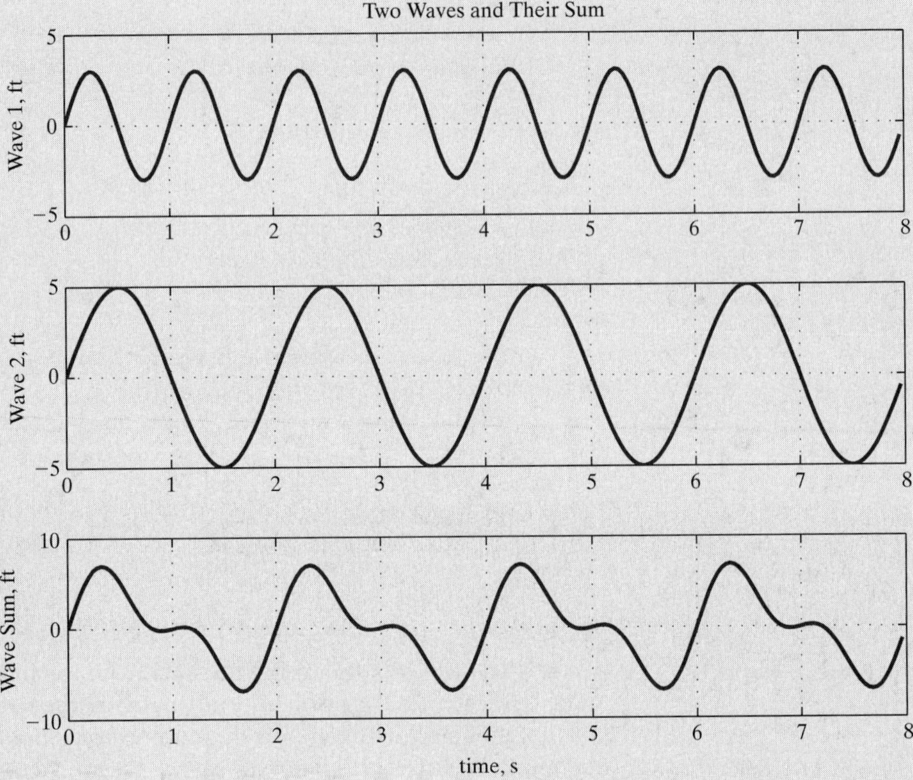

Figure 4.18
Plots of s1, s3, and their sum.

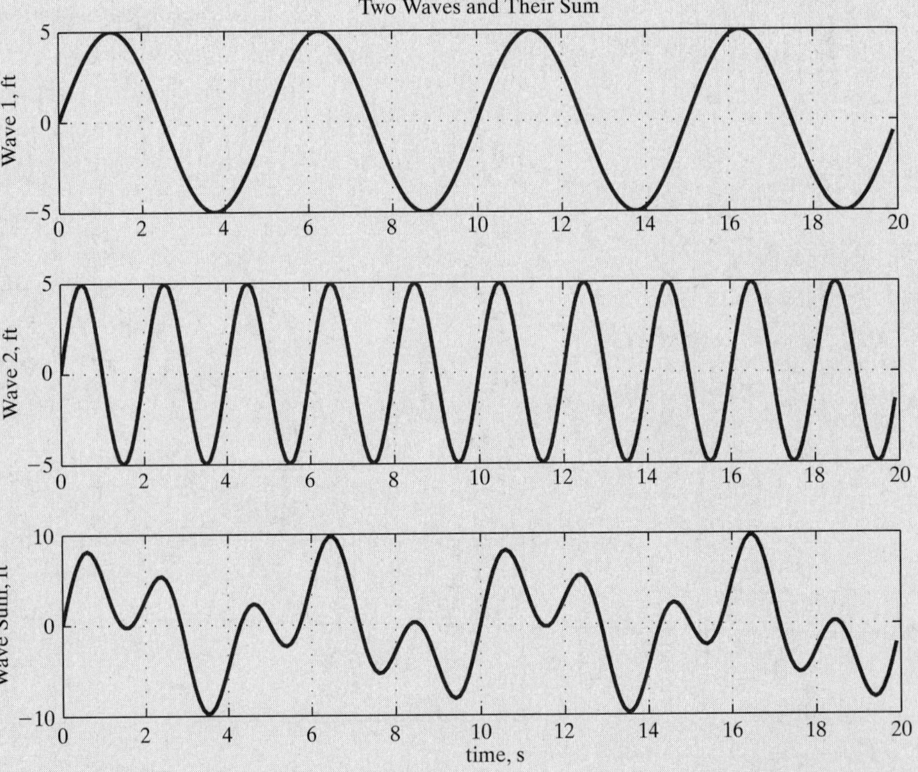

Figure 4.19
Plots of s2, s3, and their sum.

4.2 THREE-DIMENSIONAL PLOTS

MATLAB offers a variety of **three-dimensional plot** commands. Several of these formats are discussed in this section.

4.2.1 Three-Dimensional Line Plot

The **plot3** function is similar to the **plot** function, except that it accepts data in three dimensions. Instead of just providing **x** and **y** vectors, the user must also provide a **z** vector. These ordered "triples" are then plotted in three-spaces and connected with straight lines. For example, the statement below generates the plot shown in Figure 4.20:

```
x = linspace(0,10*pi,100);
y = cos(x);
z = sin(x);
plot3(x,y,z),grid,title('A Spring'),
    xlabel('angle'),ylabel('cos(x)'),zlabel('sin(x)')
```

Note that the title, labels, and grid are added in the usual way, with the addition of **zlabel** for the *z*-axis. The coordinate system used with **plot3** is oriented using the right-handed coordinate system often used in engineering.

4.2.2 Mesh Plots

A **mesh plot** is a rectilinear grid formed from a set of three-dimensional data points. They are often generated with a single two-dimensional ($n \times m$) matrix. To illustrate, assume that the values in a matrix represents the **z** values in the plot.

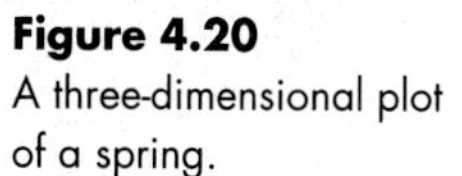

Figure 4.20
A three-dimensional plot of a spring.

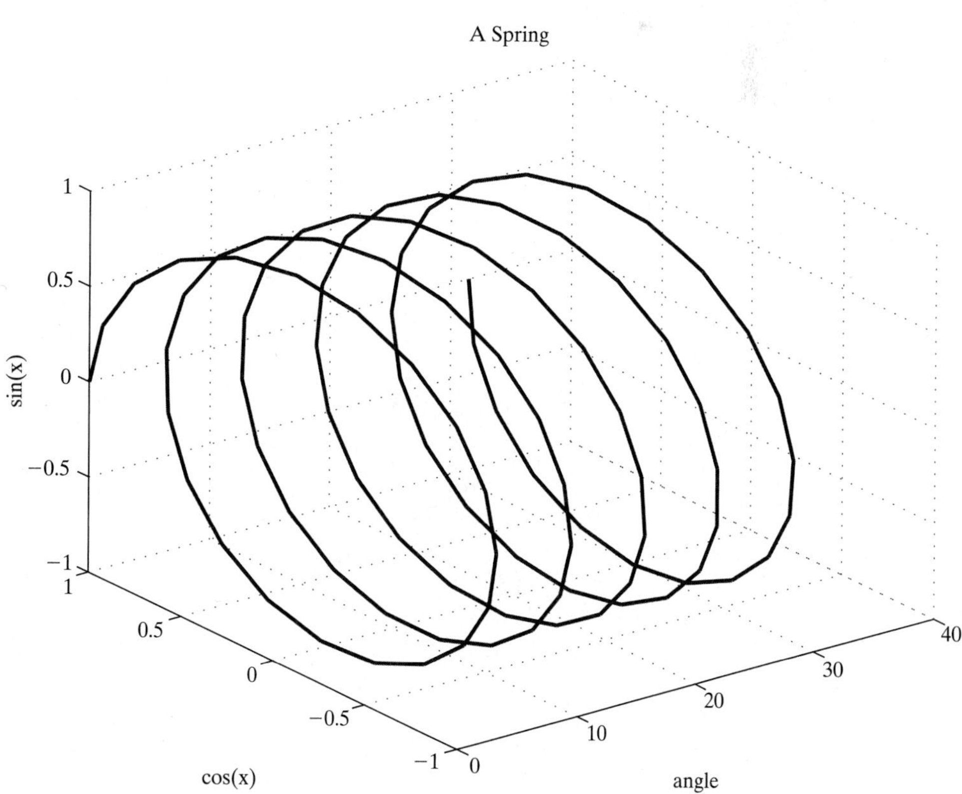

The **x** and **y** values are based on the matrix dimensions. For example, the following commands generate the graph shown in Figure 4.21:

```
z = [1, 2, 3, 4, 5, 6, 7, 8, 9,10;
     2, 4, 6, 8,10,12,14,16,18,20;
     3, 4, 5, 6, 7, 8, 9,10,11,12];
mesh(z),title('Simple Mesh Plot'),xlabel('x-axis'),
     ylabel('y-axis'),zlabel('z-axis')
```

The graph created is a mesh formed by connecting the points defined in **z** into the rectilinear grid. Notice that the *x*-axis goes from 1 to 10, with each point coordinate corresponding to a row; and *y* goes from 1 to 3, with each point corresponding to a column.

The **mesh** function can also be used with three arguments, **mesh(x,y,z)**, where each is a two-dimensional matrix. In this case, **x** is a list of *x*-coordinates, **y** is a list of *y*-coordinates, and **z** is a list of *z*-coordinates. All three matrices must be the same size, since together they represent a list of ordered triples defining the plotting points:

The following statements create a two-dimensional **z** matrix, with 10 rows and 3 columns, but it only creates one-dimensional **x** and **y** matrices.

```
z = [1, 2, 3, 4, 5, 6, 7, 8, 9,10;
     2, 4, 6, 8, 10, 12,14,16,18,20;
     3, 4, 5, 6, 7, 8, 9,10,11,12];
x = linspace(1,50,10);
y = linspace(500,1000,3);
```

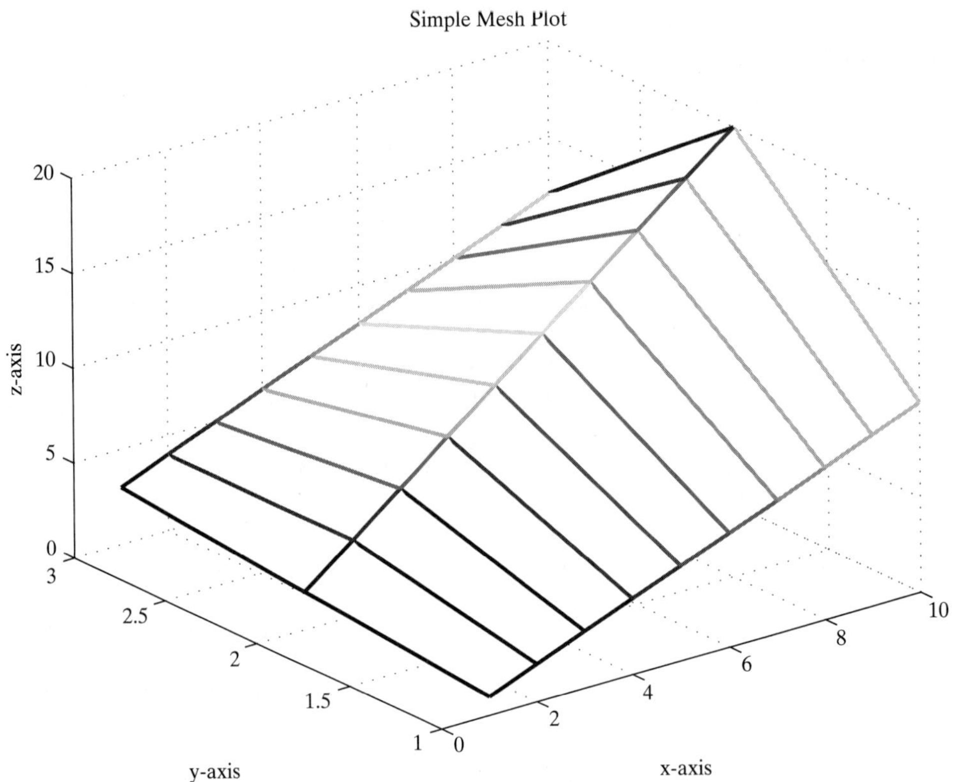

Figure 4.21
Simple mesh created with a single two-dimensional matrix.

The **x** matrix is one row by 10 columns, and the **y** matrix is one row by three columns. We can then use the **meshgrid** function to create two matrices that are the same size, three rows by 10 columns. These next commands set the output to decimal positions and then print the values of the two matrices generated by the **meshgrid** command:

```
format bank
[new_x,new_y] = meshgrid(x,y)
```

The output is:

```
new_x =
Columns 1 through 6
    1.00      6.44     11.89     17.33     22.78     28.22
    1.00      6.44     11.89     17.33     22.78     28.22
    1.00      6.44     11.89     17.33     22.78     28.22
Columns 7 through 10
    33.67       39.11       44.56       50.00
    33.67       39.11       44.56       50.00
    33.67       39.11       44.56       50.00
new_y =
Columns 1 through 6
     500.00     500.00     500.00     500.00     500.00     500.00
     750.00     750.00     750.00     750.00     750.00     750.00
    1000.00    1000.00    1000.00    1000.00    1000.00    1000.00
Columns 7 through 10
     500.00     500.00     500.00     500.00
     750.00     750.00     750.00     750.00
    1000.00    1000.00    1000.00    1000.00
```

Notice that in the case of **new_x** all the rows are the same, whereas in the case of **new_y** all the columns are the same. The creation of these new matrices provides us with appropriate input to the **mesh** function

```
mesh(new_x,new_y,z)
```

which then generates the upper left plot in Figure 4.22. The **meshgrid** function is also useful in calculations with multiple variables.

4.2.3 Surface Plots

Surface plots are similar to mesh plots, but the function **surf** creates a three-dimensional colored **surface plot** instead of a mesh. The colors vary depending on the value of **z**. The **surf** command expects the same input as **mesh**: either a single input such as **surf(z)**, in which case it uses the row and column indices as *x*- and *y*-coordinates; or three two-dimensional matrices. The plot in the upper right of Figure 4.22 was generated using the same commands as those used to generate the upper left plot except that **surf** replaced **mesh**.

The **shading** scheme for surface plots is controlled with the **shading** command. The default, shown in the upper right plot of Figure 4.22 is "faceted flat." Interpolated shading can create interesting effects as shown in the plot in the lower left of Figure 4.22 which was created with this command:

```
surf(new_x,new_y,z),shading interp
```

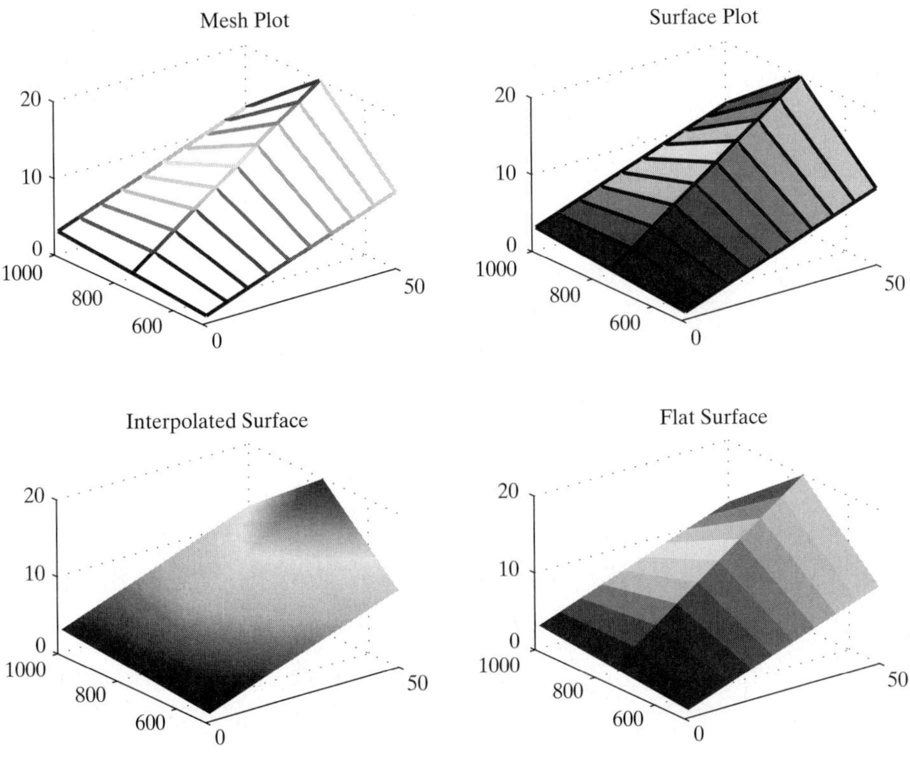

Figure 4.22
Mesh and surface plots created using three input arguments.

Flat shading is generated when we use

```
surf(new_x,new_y,z),shading flat
```

as shown in the lower right plot in Figure 4.22.

The color scheme used in surface plots can be controlled using the **colormap** function. For example,

```
colormap(gray)
```

forces a gray-scale representation for surface plots. This may be more appropriate if you will be making black and white copies of your plots. Other **colormaps** available include:

```
autumn          flag
spring          hot
summer          hsv
winter          lines
jet (default)   pink
bone            prism
colorcube       vga
cool            white
copper
```

Use the **help graph3d** command to see a description of these options.

4.2.4 Contour Plots

Contour plots are two-dimensional representations of three-dimensional surfaces. Maps often represent elevations with contours. MATLAB offers an example function called **peaks** that generates an interesting three-dimensional shape, as seen in Figure 4.23. The top plots were generated using these commands:

```
subplot(2,2,1),surf(peaks),
   title('Surface Plot'),
subplot(2,2,2),surf(peaks),shading interp,
   title('Surface Plot with Shading'),
```

The **contour** command was used to create the lower left plot and the **surfc** command was used to create the lower right plot:

```
subplot(2,2,3),contour(peaks),
   title('Contour Plot'),
subplot(2,2,4),surfc(peaks),
   title('Combination Surface and Contour Plot')
```

Additional options for using all of the three-dimensional plotting functions are included in the Help browser.

4.3 EDITING PLOTS FROM THE FIGURE WINDOW

In addition to controlling the way your plots look by using MATLAB commands, you can also edit a plot once you have created it. The plot in Figure 4.24 was created using the **sphere** command, which is one of several example functions, like **peaks**, used for demonstrating plotting.

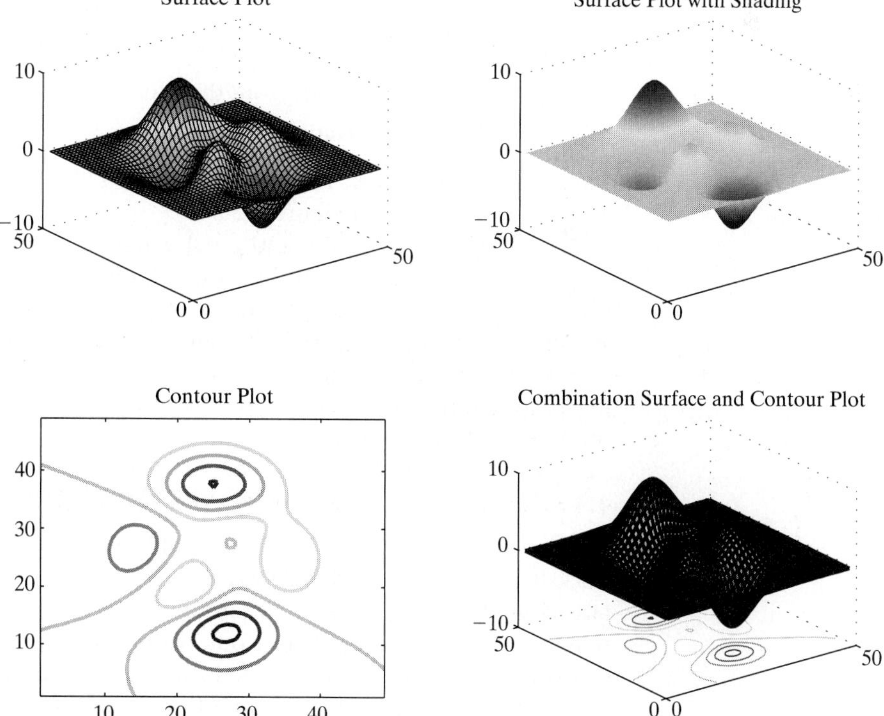

Figure 4.23
Surface and contour plots of the **peaks** function.

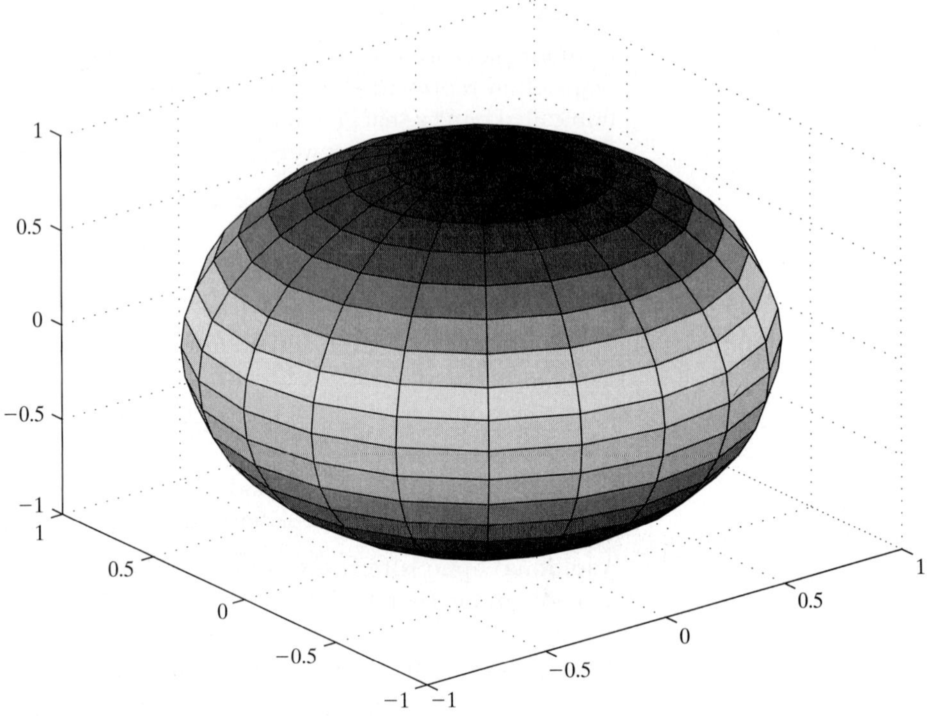

Figure 4.24
A plot of a sphere.

To annotate a graph, select the **Insert** menu in the figure window. Notice that you can insert items such as labels, titles, legends, and text boxes, all by using this menu. The **Tools menu** allows you to change the way the plot looks by performing functions such as zooming in or out, or rotating a 3D plot. The plot in Figure 4.24 does not really look like a sphere. We can adjust the shape editing this plot:

Select **Edit → Axes Properties** from the menu tool bar.

Choose **More Properties** from the box on the lower right.

Select **Data Aspect Ratio manual**.

The **Tools Menu** screen is shown in Figure 4.25, and the plot generated by these instructions is shown in Figure 4.26.

Similarly, labels, a title, and a color bar can be added using the **Insert menu** option on the menu bar. Editing your plot in this manner is more interactive and allows you to fine-tune its appearance. Once you have the plot looking the way you would like, simply go to the **Edit menu** and select **Copy Figure**. This will allow you to paste your results into a word processing document. The main problem with editing a figure interactively is that you will lose all of your improvements when you run the MATLAB program again.

4.4 CREATING PLOTS FROM THE WORKSPACE WINDOW

Another feature of MATLAB 7 is the ability to interactively create plots from the workspace window. In the workspace window, select a variable, then select the Plots tab. This tab will show a set of different plot types, including plot, bar, area, pie, hist, semilog, stem, and stairs. Select one of the plot types and MATLAB will automatically plot your data in the selected format. The interactive environment is a rich resource. You will get the most out of it by exploring and experimenting.

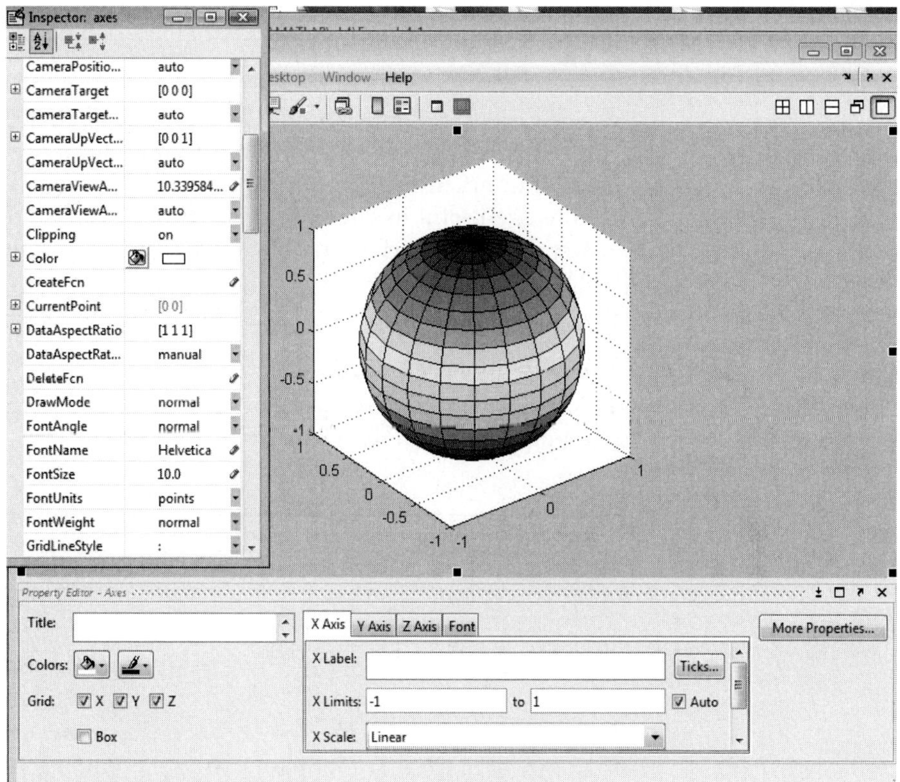

Figure 4.25
Editing from the toolbar.

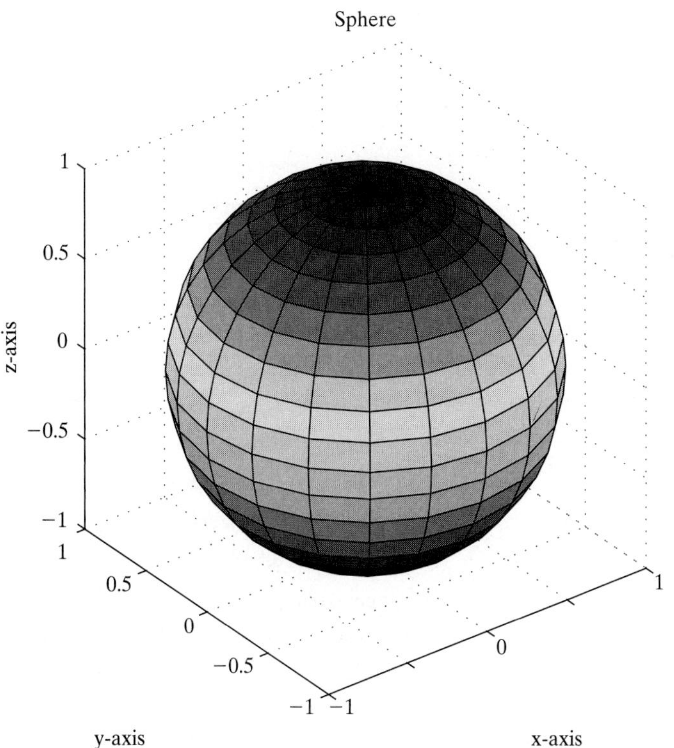

Figure 4.26
Edited plot of a sphere.

SUMMARY

A wide variety of plotting options are available in MATLAB, many of which were not covered here. Use the **help** browser to find out more about graphing options, or use the interactive graphing capability accessible from the workspace window. Graphs are useful in engineering to display results, identify trends, and pinpoint errors. No engineering graph should ever be presented without a title and axis labels. Although extremely complicated graphics are possible in MATLAB, the simple *x-y* plot is the one most commonly used by engineers.

MATLAB Summary

This MATLAB summary lists all of the special characters, commands, and functions that were defined in this chapter:

Special Characters					
Line Type	**Indicator**	**Point Type**	**Indicator**	**Color**	**Indicator**
solid	-	point	.	blue	b
dotted	:	circle	o	green	g
dash-dot	-.	x-mark	x	red	r
dashed	--	plus	+	cyan	c
		star	*	magenta	m
		square	s	yellow	y
		diamond	d	black	k
		triangle down	v		
		triangle up	^		
		triangle left	<		
		triangle right	>		
		pentagram	p		
		hexagram	h		

Commands and Functions	
`axis`	freezes the current axis scaling for subsequent plots or specifies the axis dimensions
`bar`	generates a bar graph
`bar3`	generates a three-dimensional bar graph
`barh`	generates a horizontal bar graph
`bar3h`	generates a horizontal three-dimensional bar graph
`colormap`	color scheme used in surface plots
`figure`	opens a new figure window
`grid`	adds a grid to the current graph in the current figure

MEASUREMENT DATA

The vector **G** represents the distribution of measurements:

G = [68, 83, 61, 70, 75, 82, 57, 51, 76, 85, 62, 71, 96, 78, 76, 68, 72, 75, 83, 93]

7. Use MATLAB to sort the data and create a bar graph of the scores.
8. Create a histogram of the scores.

QUALITY MEASUREMENTS

The measurement data used in Problems 7 and 8 can also be used to give qualitative assessment. These scores represent 2 As, 4 Bs, 8 Cs, 4 Ds, and 2 Fs.

9. Create a pie chart of this distribution. Add a legend of the assessments (A, B, C, etc.).
10. Use the menu text option to add a text box to each pie slice (instead of a legend).
11. Create a three-dimensional pie chart of the same data.

FUNCTIONAL DATA

Create a vector of x values from 0 to 20π with a spacing of $\pi/100$, where

$$y = x \cdot \sin(x)$$

$$z = x \cdot \cos(x)$$

12. Create an x-y plot of x and y.
13. Create a polar plot of x and y.
14. Create a three-dimensional line plot of x, y, and z. Be sure to add a title and labels.
15. Figure out how to adjust your input to **`plot3`** in Problem 14, to create a graph that looks like a tornado.

PEAKS

Use the following information for Problems 16 through 20. The equation used in the **`peaks`** function is

```
z = 3*(1-x).^2.*exp(-(x.^2) - (y+1).^2) ...
    - 10*(x/5 - x.^3 - y.^5).*exp(-x.^2-y.^2) ...
    - 1/3*exp(-(x+1).^2 - y.^2);
```

where **`z`** is a two-dimensional matrix determined by the values of **`x`** and **`y`**. Because **`z`** is a two-dimensional matrix, both **`x`** and **`y`** must be two-dimensional matrices. Create a vector **`x`**, from -3 to $+3$ with 100 values. Create a vector **`y`**, from -3 to $+3$ with 100 values. Use the **`meshgrid`** function to map **`x`** and **`y`** into two-dimensional matrices, and then use them to calculate **`z`**.

16. Use the **`mesh`** plotting function to create a three-dimensional plot of **`z`**.
17. Use the **`surf`** plotting function to create a three-dimensional plot of **`z`**. Compare the results you generate using a single input (**`z`**), or inputs for all three dimensions (**`x,y,z`**).
18. Modify your surface plot with interpolated shading. Try using different colormaps.
19. Generate a contour plot of **`z`**.
20. Generate a combination surface, contour plot of **`z`**.

CHAPTER

5 Control Structures

Objectives

After reading this chapter, you should be able to

- use relational and logical operators in conditions that are true or false,
- use the conditions with the `find` function to select values from vectors and matrices, and
- implement `for` loops and `while` loops in MATLAB.

ENGINEERING ACHIEVEMENT: SIGNAL PROCESSING

Computer algorithms for word recognition are complicated algorithms that work best when the speech signals are "clean." However, when speech signals are collected by microphones, the background noise is also collected. Therefore, preprocessing steps are often used to remove some of the background noise before attempting to identify the words in the speech signals. These preprocessing steps may require a number of operations that fall into the area of signal processing, such as analyzing the characteristics of a signal, decomposing a signal into sums of other signals, coding a signal in a form that is easy to transmit across a communication channel, and extracting information from a signal. Some of the functions commonly used in signal processing are used in examples in this chapter.

5.1 RELATIONAL AND LOGICAL OPERATORS

The MATLAB programs that we have developed have been based on performing mathematical operations, using functions, and printing or plotting the results of these computations. These programs have had a **sequential structure**; that is, the commands were executed one after another in a serial fashion. There are two other types of structures called selection structures and repetition structures. Definitions of these

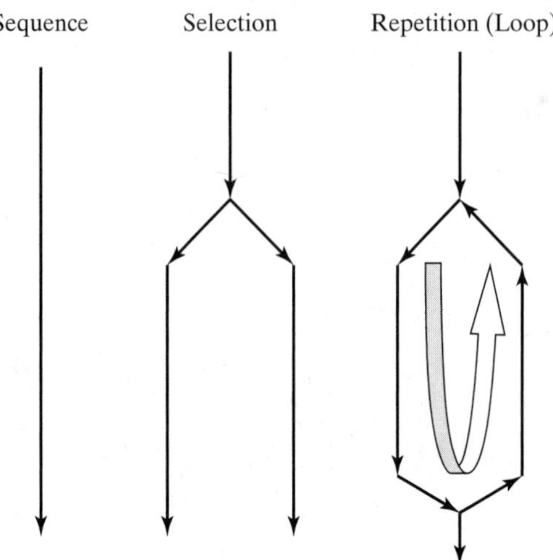

Figure 5.1
Sequence, selection, and loop control structures.

three **control structures** are now given; Figure 5.1 shows a graphical depiction of these structures.

- Sequences are lists of commands that are executed one after another.
- A **selection structure** allows the programmer to execute one command (or group of commands) if some criteria is true, and a second set of commands if the criteria is false. A selection statement provides the means of choosing between these paths based on a **logical condition**. The **conditions** that are evaluated often contain **relational** and **logical** operators or functions.
- A **repetition structure**, or **loop**, causes a group of statements to be executed zero, one, or more times. The number of times a loop is executed depends on either a counter or the evaluation of a logical condition.

Before presenting the MATLAB commands for selection statements and loops, we first need to define **relational operators** and **logical operators**. MATLAB has six relational operators for comparing two matrices of equal size, as shown in Table 5.1. Comparisons are either true or false, and most computer languages (including MATLAB) use the number 1 for true and 0 for false.

Assume we define two scalars as shown:

```
x = 5;
y = 1;
```

and use a relational operator such as < (for less than) with the two variables, as in:

```
x<y
```

Table 5.1 Relational Operators

Relational Operator	Interpretation
<	less than
<=	less than or equal to
>	greater than
>=	greater than or equal to
==	equal to
~=	not equal to

This new expression (**x<y**) is evaluated to be either true or false. In this case, **x** is not less than **y**, so MATLAB responds

```
ans =
      0
```

indicating the comparison was not true. MATLAB uses this answer in selection statements and in repetition structures to make decisions.

Of course, variables in MATLAB usually represent entire matrices. If we redefine **x** and **y**, we can see how MATLAB handles comparisons between matrices:

```
x = 1:5;
y = x-4;
x<y
ans =
      0    0    0    0    0
```

MATLAB compares corresponding elements and creates an answer matrix of zeros and ones. In the previous example, **x** was greater than **y** for every element comparison, so every comparison was false, and the answer was a string of zeros. Consider these statements:

```
x = [ 1, 2, 3, 4, 5];
y = [-2, 0, 2, 4, 6];
x<y
ans =
      0    0    0    0    1
```

The result tells us that the comparison was false for the first four elements, but true for the last. In order for MATLAB to decide that a comparison is true for an entire matrix, it must be true for every element in the matrix. In other words, all of the results must be one.

MATLAB also allows us to combine comparisons with logical operators **and**, **not**, and **or**, as shown in Table 5.2.

Consider the following commands and corresponding output:

```
x = [ 1, 2, 3, 4, 5];
y = [-2, 0, 2, 4, 6];
z = [ 8, 8, 8, 8, 8];
z>x & z>y
ans =
      1    1    1    1    1
```

Since **z** is greater than both **x** and **y** for every element, the condition is true for each corresponding set of values. The statement

```
x>y | x>z
```

Table 5.2 Logical Operators

Logical Operator	Interpretation
&	and
~	not
|	or

is read as "**x** is greater than **y** or **x** is greater than **z**" and would return an answer as the following:

```
ans =
      1   1   1   0   0
```

This result is interpreted to mean that the condition is true for the first three elements and false for the last two.

These relational and logical operators are used in both selection structures and loops to determine what commands should be executed.

5.2 SELECTION STRUCTURES

MATLAB offers two kinds of selection structures: the function **find**, and a family of **if** structures.

5.2.1 The find Function

The **find** command is unique to MATLAB, and can often be used instead of both if and loop structures. It returns a vector composed of the indices of the nonzero elements of a vector **x**. Those indices can then be used in subsequent commands. The usefulness of the **find** command is best described with examples.

Assume that we have a vector **d** containing a group of distance values that represent the distances of a cable car from the nearest tower. We want to generate a vector containing velocities of the cable car at those distances. If the distance of the cable car from the tower is less than or equal to 30 ft, we use this equation to compute the velocity:

$$\text{velocity} = 0.425 + 0.00175\,\text{d}^2$$

If the cable car is farther than 30 ft from the tower, we use the following equation:

$$\text{velocity} = 0.625 + 0.12\,\text{d} - 0.00025\,\text{d}^2$$

We can use the **find** function to find the distance values greater than 30 ft and the distance values less than or equal to 30 ft. Because the **find** function identifies the subscripts for each group of values, we can compute the corresponding velocities with these statements:

```
lower = find(d<=30);
velocity(lower) = 0.425 + 0.00175*d(lower).^2;
upper = find(d>30);
velocity(upper) = 0.625 + 0.12*d(upper) - 0.00025*d(upper).^2;
```

If all the values of **d** are less than or equal to 30, the vector **upper** will be an empty vector, and the reference to **d(upper)** and **velocity(upper)** will not cause any values to change.

Our next example assumes that you have a list of temperatures measured in a manufacturing process. If the temperature is less than 95°F, the items produced will be faulty. Assume that the temperatures for a set of items are the following:

```
temp = [100,98,94,101,93];
```

Use the **find** function to determine which items are faulty:

```
find(temp<95)
```

returns a vector of element numbers:

```
ans =
      3   5
```

which tells us that items 3 and 5 will be faulty. MATLAB first evaluated **temp<95**, which resulted in a vector of zeros and ones. We can see this by typing the comparison into MATLAB:

```
temp<95
```

which returns a vector indicating when the comparison was true (1) and when it was false (0):

```
ans =
      0   0   1   0   1
```

The **find** command identified the elements for which the comparison was true (where the **ans** vector reported ones).

It is also sometimes useful to name these element lists. For example,

```
faulty = find(temp<95);
pass = find(temp>=95);
```

makes it possible to create a results table:

```
failtable = [faulty',temp(faulty)']
```

which returns a table of elements and the corresponding temperatures:

```
failtable =
              3   94
              5   93
```

When the **find** command is used with a two-dimensional matrix, a single element number is returned. As discussed before, MATLAB is a column dominant language and considers two-dimensional matrices as one long list of numbers. Just as **fprintf** works down one column at a time, the **find** function uses an element numbering scheme that works down each column one at a time. For example, consider a 10-by-3 matrix. The element numbers are shown in Figure 5.2.

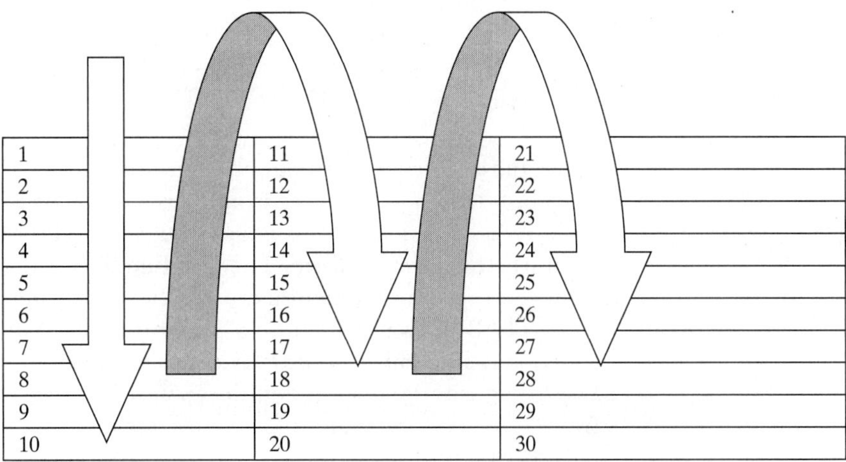

Figure 5.2
Element numbering sequence for a 10 × 3 matrix.

An alternate way to use **find** returns the row and column designation of an element:

```
[row, column] = find(expression)
```

For example, consider the following two-dimensional matrix, and use **find** to determine the location of all elements greater than 9:

```
x = [1,2,3; 10,5,1; 3,12,2; 8,3,1]
element = find(x>9)
[row,column] = find(x>9)
```

returns

```
x =
    1   2   3
   10   5   1
    3  12   2
    8   3   1
element =
         2
         7
row =
     2
     3
column =
        1
        2
```

Notice that the numbers 10 and 12 are the only two values greater than 9. By counting down the columns, we see that they are elements 2 and 7 respectively. Using the alternative designation, 10 is in row 2, column 1, and 12 is in row 3, column 2.

EXAMPLE 5.1

THE SINC FUNCTION

The sinc function is used in many engineering applications, but especially in signal processing applications. Unfortunately, there are two widely accepted definitions for this function:

$$f_1(x) = \frac{\sin(\pi x)}{\pi x} \text{ and } f_2(x) = \frac{\sin x}{x}$$

Both of these functions have an indeterminate form of $0/0$ when x is equal to 0. In this case, L'Hôpital's theorem from calculus can be used to prove that both functions are equal to 1 when x is equal to 0. For values of x not equal to 0, these two functions have a similar form. The first function, $f_1(x)$, crosses the x-axis when x is an integer; the second function crosses the x-axis when x is a multiple of π.

MATLAB does not include a sinc function. Assume that you would like to define a function called **sinc1** that uses the first definition, and a function called **sinc2** that uses the second definition. Test your functions by calculating values of x from -5π to $+5\pi$ and plotting the results.

SOLUTION

1. Problem Statement

Create and test functions for the two definitions of the sinc function:

$$f_1(x) = \frac{\sin(\pi x)}{\pi x} \text{ and } f_2(x) = \frac{\sin x}{x}$$

2. Input/Output Description

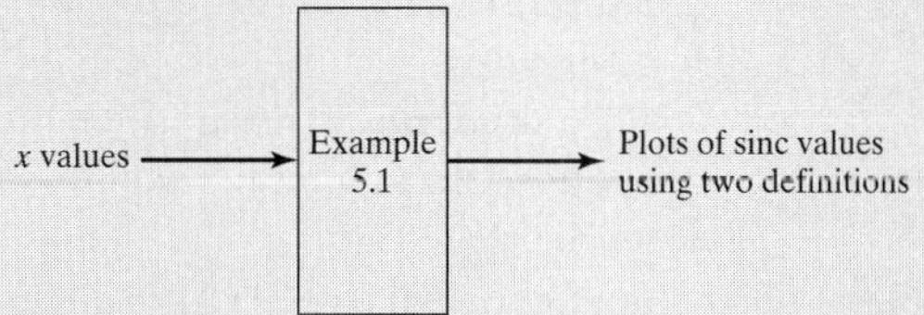

3. Hand Example

The following table computes a few values of these sinc functions

x	sin(πx)/(πx)	sin(x)/(x)
0	0/0 = 1	0/0 = 1
1	0	sin(1)/1 = 0.8415
2	0	sin(2)/2 = 0.4546
3	0	sin(3)/3 = 0.0470

4. MATLAB Solution

First, create the functions **sinc1** and **sinc2**:

```
%--------------------------------------------------------------
function output = sinc1(x)
%  This function calculates the value of sinc
%  using the definition of sin(pi*x)/(pi*x).
%
%  Determine the elements in x that are close to zero.
set1 = find(abs(x)<0.0001);
%
%  Set those elements in the output array to 1.
output(set1) = 1;
%
%  Determine elements in x that are not close to 0.
set2 = find(abs(x)>=0.0001);
%
%  Calculate sin(pi*x)/(pi*x) for the elements not close to 0.
output(set2) = sin(pi*x(set2))./(pi*x(set2));
%--------------------------------------------------------------
```

(*continued*)

```
%-------------------------------------------------------------
function output = sinc2(x)
%  This function calculates the value of sinc
%  using the definition of sin(x)/x.
%
% Determine the elements in x that are close to zero.
set1 = find(abs(x)<0.0001);
%
% Set those elements in the output array to 1.
output(set1) = 1;
%
% Determine elements in x that are not close to 0.
set2 = find(abs(x)>=0.0001);
%
%  Calculate sin(x)/x for the elements not close to 0.
output(set2) = sin(x(set2))./x(set2);
%-------------------------------------------------------------
```

Once we have created the functions in M-files, we can test them in the command window using the inputs from the hand example:

```
sinc1(0)
ans =
     1
sinc1(1)
ans =
     3.8982e-17
sinc2(0)
ans =
     1
sinc2(1)
ans =
     0.8415
```

When we compare the results with the hand example, we see that the answers match. (Although you will see when you run this example that instead of zero, you will get very small values in some cases.)

5. Testing

We now use the functions in our M-file solution to Example 5.1 with confidence.

```
%-------------------------------------------------------------
%  Example 5_1 This program plots the sinc function
%  using two definitions for the sinc function.
%
clear, clc
%
%  Define an array of angles
x = -5*pi:pi/100:5*pi;
%
%  Calculate sinc1 and sinc2.
```

```
y1 = sinc1(x);
y2 = sinc2(x);
%
%   Create the plot.
subplot(2,1,1),plot(x,y1),
   title('Two Definitions for sinc Function'),
   ylabel('sin(pi*x)/(pi*x)'),grid,
   axis([-5*pi,5*pi,-0.5,1.5]),
subplot(2,1,2),plot(x,y2),xlabel('angle, radians'),
   ylabel('sin(x)/x'),grid,
   axis([-5*pi,5*pi,-0.5,1.5])
%-----------------------------------------------------------------
```

This program generates the plot shown in Figure 5.3. The plots indicate that the function is working properly. Testing the functions with one value at a time validated its answers for a scalar input; however, the program that generated the plots sent a vector argument to the functions. The plots confirm that it also performs properly with vector input.

If you have trouble understanding how this function works, remove the semicolons that are suppressing the output, and run the program. Understanding the output from each line will help you understand the program logic better.

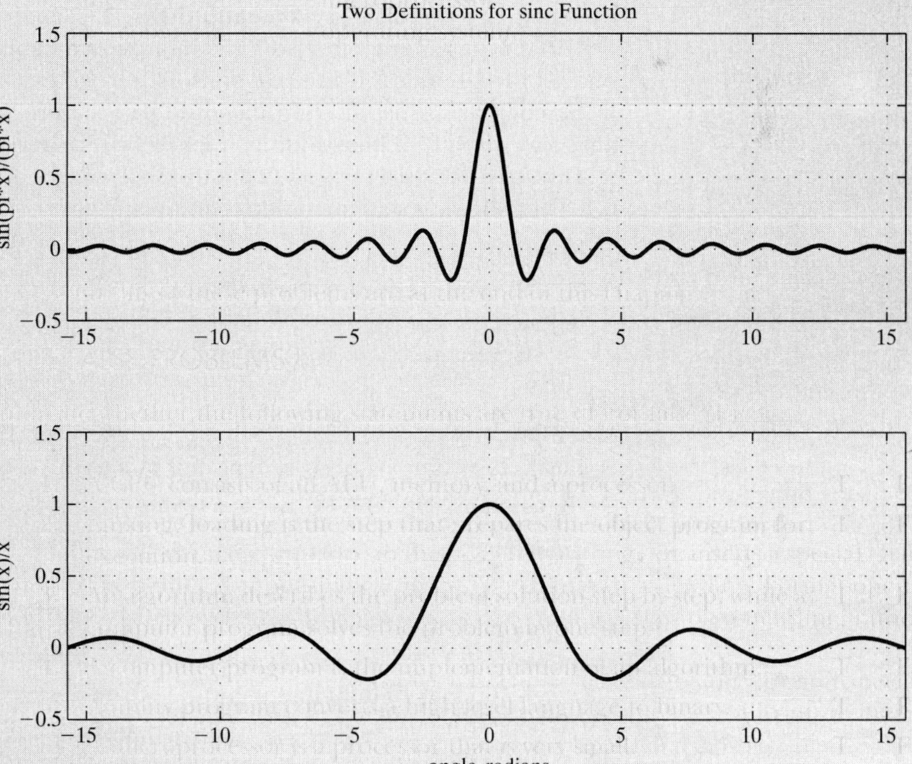

Figure 5.3
Plots of alternative definitions for the sinc function.

5.2.2 The Family of `if` Statements

There are situations when the **find** statement does not do what we need to do. In these cases, we can use the **if** statement which has the following form:

```
if comparison
      statements
end
```

If the logical expression (the comparison) is true, the statements between the **if** statement and the **end** statement are executed. If the logical expression is false, program control jumps immediately to the statement following the **end** statement. It is good programming practice to indent the statements within an **if** structure for readability. For example,

```
if G<50
   count = count + 1;
   disp(G);
end
```

This statement (from **if** to **end**) is easy to interpret if **G** is a scalar. For example, if **G** has a value of 25, then **count** is incremented by 1 and **G** is displayed on the screen. However, if **G** is not a scalar, then the **if** statement considers the comparison true only if it is true for every element. If **G** is defined as:

```
G = 0:10:80;
```

then the comparison is false, and the statements inside the **if** statement are not executed. In general, **if** statements work best when dealing with scalars.

An **if** statement can also contain an **else** clause that allows us to execute one set of statements if the comparison is true, and a different set of statements if the comparison is false. To illustrate this feature, assume that we have a variable **interval**. If the value of **interval** is less than 1, we set the value of **x_increment** to **interval/10**; otherwise, we set the value of **x_increment** to 0.1. The following statement performs these steps:

```
if interval < 1
   x_increment = interval/10;
else
   x_increment = 0.1;
end
```

When **interval** is a scalar, this is easy to interpret. However, when **interval** is a matrix, the comparison is only true if it is true for every element in the matrix. So, if

```
interval = 0:0.5:2;
```

the elements in the matrix are not all less than 1. Therefore, MATLAB skips to the **else** portion of the statement, and all values in the **x_increment** vector are set equal to 0.1. Again, **if/else** statements are probably best used with scalars, although you may find limited use with vectors.

When we need more levels of **if-else** statements, it may be difficult to determine which logical expressions must be true (or false) to execute each set of statements. In these cases, the **elseif** clause is often used to clarify the program logic, as illustrated in the following statement:

```
if temperature>100
   disp('Too hot-equipment malfunctioning')
elseif temperature>90
   disp('Normal operating temperature')
elseif temperature>50
   disp('Temperature below desired operating range')
else
   disp('Too cold-turn off equipment')
end
```

In this example, temperatures above 90°F and below or equal to 100°F are in the normal operating range. Temperatures outside of this range generate an appropriate message. Notice that a temperature of 101°F does not trigger all of the responses; it only displays the first message. Also notice that the final **else** does not require a comparison. In order for the computation to reach the final **else**, the temperature must be less than or equal to 50°F. Again, this structure is easy to interpret if **temperature** is a scalar. If it is a matrix, the comparison must be true for every element in the matrix.

As before, **elseif** structures work well for scalars, but **find** is probably a better choice for matrices. Here is an example from a script file with an array of temperatures that generates a table of results in each category using the **find** statement:

```
temperature = [90,45,68,84,92,95,101];
%
set1 = find(temperature>100);
disp('Too Hot-Equipment Malfunctioning')
disp('Element, Temperature')
table1 = [set1; temperature(set1)];
fprintf('%1.0f   %8.0f \n',table1)
%
set2 = find(temperature>90 & temperature<=100);
disp('Normal Operating Temperature')
disp('Element, Temperature')
table2 = [set2; temperature(set2)];
fprintf('%1.0f   %8.0f \n',table2)
%
set3 = find(temperature>50 & temperature<=90);
table3 = [set3; temperature(set3)];
disp('Temperature Below Desired Operating Range')
disp('Element, Temperature')
fprintf('%1.0f   %8.0f \n',table3)
%
set4 = find(temperature<=50);
table4 = [set4; temperature(set4)];
disp('Too Cold-Turn Off Equipment')
disp('Element, Temperature')
fprintf('%1.0f   %8.0f \n',table4)
```

The output is:

```
Too Hot—Equipment Malfunctioning
Element, Temperature
7         101
Normal Operating Temperature
Element, Temperature
5         92
6         95
Temperature Below Desired Operating Range
Element, Temperature
1         90
3         68
4         84
Too Cold-Turn Off Equipment
Element, Temperature
2         45
```

EXAMPLE 5.2

ASSIGNING PERFORMANCE QUALITY SCORES

The **if** statement is used most effectively when the input is a scalar. Create a function to determine performance quality based on a quality score, assuming a single input to the function. The quality grades should be based on the following criteria:

Quality Grade	Score
A	>= 90 to 100
B	>= 80 and < 90
C	>= 70 and < 90
D	>= 60 and < 70
F	< 60

SOLUTION

1. Problem Statement

Determine the performance quality based on a numeric score.

2. Input/Output Description

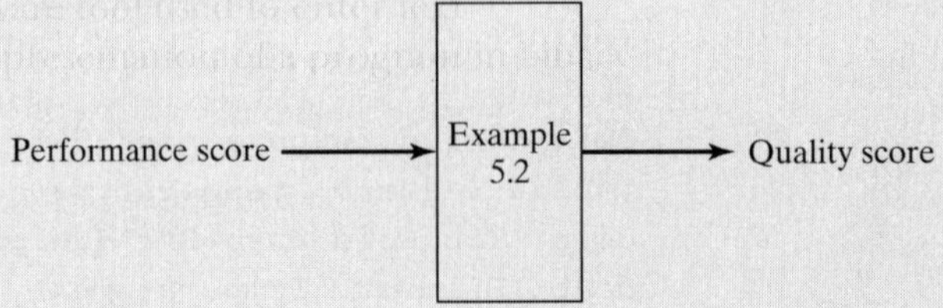

3. Hand Example

For example, a score of 85 should be a B.

4. MATLAB Solution

First, create the function:

```
%-------------------------------------------------------------
function results = quality(x)
% This function requires a scalar input
%
if(x>=90)
   results = 'A';
elseif(x>=80)
   results = 'B';
elseif(x>=70)
   results = 'C';
elseif(x>=60)
   results = 'D';
else
   results = 'F';
end
%-------------------------------------------------------------
```

5. Testing

We can now test the function in the command window:

```
quality(25)
ans =
     F
quality(80)
ans =
     B
quality(-52)
ans =
     F
```

Notice that, although the function seems to work properly, it returns quality scores for values over 100 and less than 0. You can now go back and add the logic to exclude those values using nested **if** structures:

```
%-------------------------------------------------------------
function results = grade(x)
%  This function converts a scalar input to a score
%  that is a quality score.
if(x>=0 & x<=100)
   if(x>=90)
      results = 'A';
   elseif(x>=80)
      results = 'B';
   elseif(x>=70)
      results = 'C';
   elseif(x>=60)
      results = 'D';
```

(*continued*)

```
      else
         results = 'F';
      end
   else
      results = 'Illegal Input';
   end
   %-------------------------------------------------------------
```

In the command window, we can test the function again:

```
grade(-10)
ans =
     Illegal Input
grade(108)
ans =
     Illegal Input
```

This function will work well for scalars, but if you send a vector to the function, you may get some unexpected results of the vector comparisons in the function:

```
score = [95,42,83,77];
grade(score)
ans =
     F
```

5.3 LOOPS

A loop is a structure that allows you to repeat a set of statements. In general, you should avoid loops in MATLAB because they are seldom needed, and they can significantly increase the execution time of a program. If you have previous programming experience, you may be tempted to use loops extensively. Try instead to formulate a solution using **find**. However, there are occasions when loops are needed, so we give a brief introduction to **for** loops and **while** loops.

5.3.1 For Loops

In general, it is possible to use either a for or a **while** loop in any situation that requires a repetition structure. However, **for** loops are the easier choice when you know how many times you want to repeat a set of instructions. The general format is

```
for index = expression
     statements
end
```

Each time through the **for** loop, the index has the value of one of the elements in the expression matrix. This can be demonstrated with a simple **for** loop:

```
for k=1:5
   k
end
```

returns

```
k =
   1
```

```
k =
    2
k =
    3
k =
    4
k =
    5
```

The rules for writing and using a **for** loop are the following:

a. The index of a **for** loop must be a variable. Although **k** is often used as the symbol for the index, any variable name can be used. The use of **k** is strictly a style issue.
b. If the expression matrix is the empty matrix, the loop will not be executed. Control will pass to thc statcmcnt following the **end** statement.
c. If the expression matrix is a scalar, the loop will be executed one time, with the index containing the value of the scalar.
d. If the expression is a row vector, each time through the loop the index will contain the next value in the vector.
e. If the expression is a matrix, each time through the loop the index will contain the next column in the matrix. This means that the index will be a column vector.
f. Upon completion of a **for** loop, the index contains the last value used.
g. The colon operator can be used to define the expression matrix using the following format:

   ```
   for k = initial:increment:limit
   ```

h. All index variables in **nested loops** must be different variables.

In an example in the previous section, we used the **find** function to find distances greater than 30 ft and distances less than or equal to 30 ft. We then computed the corresponding velocities. Another way to perform these steps uses a **for** loop. In the following statements, the value of **k** is set to 1, and the statements inside the loop are executed as many times as there are values in the vector **d**. The value of **k** is incremented to 2, and the statements inside the loop are executed again. This continues until the value of **k** is greater than the length of the **d** vector.

```
for k=1:length(d)
   if d(k)<=30
      velocity(k) = 0.425 + 0.00175*d(k).^2;
   else
      velocity(k) = 0.625 + 0.12*d(k) = - 0.00025*d(k).^2;
   end
end
```

Although these statements perform the same operations as the previous steps using the **find** function, the solution without a loop will execute much faster.

5.3.2 While Loops

The **while** loop is a structure for repeating a set of statements as long as a specified condition is true. The general format for this control structure is:

```
while expression
      statements
end
```

If the expression is true, the statements are executed. After these statements are executed, the condition is retested. If the condition is still true, the group of statements is executed again. When the condition is false, control skips to the statement following the **end** statement. The variables modified within the loop should include the variables in the expression, or the value of the expression will never change. If the expression is always true (or is a value that is nonzero), the loop becomes an **infinite loop**. (You can exit an infinite loop by typing Ctrl c.)

We used a **for** loop to compute velocities for a cable car in the previous section. We now use a **while** loop to perform the same steps. Recall that we have one equation to use to compute velocity if the distance of the cable car is less than or equal to 30 ft from a tower; otherwise we use a different equation to compute the velocity. Here is the **while** loop solution:

```
k = 1;
while k<=length(d)
   if d(k)<=30
      velocity(k) = 0.425 + 0.00175*d(k).^2;
   else
      velocity(k) = 0.625 + 0.12*d(k) - 0.00025*d(k).^2;
   end
end
```

Although these statements perform the same operations as the previous steps using the **find** function, the solution without a loop will execute much faster.

5.3.3 Loop Timing

To demonstrate that a solution without loops is much faster to execute than one with a **for** loop or a **while** loop, consider the following statements. We generate an array of ones that has 200 rows and 200 columns. We then multiply each element in the array by pi. This computation is computed three times—once without a loop, once with a for loop, and once with a while loop. Here are the statements that use the **tic** and **toc** functions to compute the execution times of each segment of code. Use the help feature to review these two useful functions.

```
%  Compute an array of ones that is 200x200.
a = ones(200);
%
%  Determine time without a loop.
tic
b = a*pi;
toc
%
%  Determine time with a for loop.
tic
for k=1:length(a(:))
   b(k) = a(k)*pi;
end
toc
%
%  Determine time for while loop.
tic
k = 1
```

```
while k<=length(a(:))
   b(k) = a(k)*pi;
   k = k+1;
end
toc
```

The output from these statements is the following:

```
Elapsed time is 0.000076 seconds.
Elapsed time is 0.003440 seconds.
Elapsed time is 0.050960 seconds.
```

The solution with a for loop takes over 45 times more computer time than the solution without a loop; the solution with a while loop takes over 670 times more computer time than the solution without a loop.

SUMMARY

In this chapter, we expanded our set of programming tools through the addition of relational and conditional operators. These operators allow us to describe conditions that can be evaluated to be true or false. The **find** command uses conditions to select elements in matrices; this command is one of the most powerful commands in MATLAB. We also introduced **if** loops, **for** loops, and **while** loops.

MATLAB Summary

This MATLAB summary lists and briefly describes all of the special characters, commands, and functions that were defined in this chapter:

Special Constants	
<	less than
<=	less than or equal to
>	greater than
>=	greater than or equal to
==	equal to
~=	not equal to
&	and
|	or
~	not

Commands and Functions	
`else`	defines the path if the result of an `if` statement is false
`elseif`	defines the path if the result of one condition is false and a second is true
`end`	identifies the end of a control structure
`find`	determines which elements in a matrix meet the input criteria
`for`	generates a loop structure
`if`	tests a logical expression
`tic`	starts a timing sequence
`toc`	stops a timing sequence
`while`	generates a loop structure

KEY TERMS

condition
control structure
logical condition
logical operator
loop
nested loops
relational operator
repetition structure
selection structure
sequential structure

PROBLEMS

DISTANCES TO THE HORIZON

The distance to the horizon increases as you climb a mountain (or a hill). The expression

$$d = \sqrt{2rh + h^2}$$

where

d = distance to the horizon,
r = radius of the Earth, and
h = height of the hill,

can be used to calculate that distance. The distance depends on how high the hill is and the radius of the Earth. Of course, on other planets the radius is different. For example, the Earth's diameter is 7,926 miles and Mars' diameter is 4,217 miles.

1. Create a MATLAB program to find the distance in miles to the horizon both on Earth and on Mars for hills from 0 to 10,000 ft, in increments of 500 ft. Remember to use consistent units in your calculations. Use the **meshgrid** function to solve this problem. Report your results in a table. Each column should represent a different planet, and each row should represent a different hill height. Be sure to provide a title for your table and column headings. Use **disp** for the title and headings; use **fprintf** for the table values.
2. Create a function called **distance** to find the distance to the horizon. Your function should accept two input vectors, **radius** and **height**, and should return a table similar to the one in Problem 1. Use the results of Problem 1 to validate your calculations.

CURRENCY CONVERSIONS

Use your favorite Internet search engine to identify recent currency conversions for British pounds sterling, Japanese yen, and the European euro to U.S. dollars. Use the conversion equations to create the following tables. Use the **disp** and **fprintf** commands in your solution, which should include a title, column labels, and formatted output.

3. Generate a table of conversions from yen to dollars. Start the yen column at ¥5 and increment by ¥5. Print 25 lines in the table.
4. Generate a table of conversions from the euro to dollars. Start the euro column at €1 and increment by €2. Print 30 lines in the table.
5. Generate a table with four columns. The first should contain dollars, the second the equivalent number of euros, the third the equivalent number of pounds, and the fourth the equivalent number of yen. The first column of the table should start with 1 and go through 25 in increments of 5.

TEMPERATURE CONVERSIONS

This set of problems requires you to generate temperature conversion tables. Use the following equations, which describe the relationships between temperatures in degrees Fahrenheit (T_F), degrees Celsius (T_C), degrees Kelvin (T_K), and degrees Rankine (T_R), respectively:

$$T_F = T_R - 459.67^\circ\mathrm{R}$$

$$T_F = \frac{9}{5}T_C + 32^\circ\mathrm{F}$$

$$T_R = \frac{9}{5}T_K$$

You will need to rearrange these expressions to solve some of the problems.

6. Generate a table with the conversions from Fahrenheit to Kelvin for values from 0°F to 200°F. Allow the user to enter the increments in degrees F between lines.
7. Generate a table with the conversions from Celsius to Rankine. Allow the user to enter the starting temperature and increment between lines. Print 25 lines in the table.
8. Generate a table with the conversions from Celsius to Fahrenheit. Allow the user to enter the starting temperature, the increment between lines, and the number of lines for the table.

ROCKET TRAJECTORY

Suppose a small rocket is being designed to make wind shear measurements in the vicinity of thunderstorms. The height of the rocket can be represented by the following equation:

$$\text{height} = 2.13t^2 - 0.0013t^4 + 0.000034t^{4.751}$$

9. Create a function called **`height`** that accepts time as an input and returns the height of the rocket. Use the function in your solutions for the next two problems.
10. Compute, print, and plot the time and height of the rocket from the time it launches until it hits the ground, in increments of 2 seconds. If the rocket has not hit the ground within 100 seconds, print values only up through 100 seconds. (Use the function from Problem 9.)
11. Modify the steps in Problem 10 so that, instead of a table, the program prints the time at which the rocket begins to fall back to the ground and the time at which it hits the ground (when the elevation becomes negative).

SUTURE PACKAGING

Sutures are strands or fibers used to sew living tissue together after an injury or an operation. Packages of sutures must be sealed carefully before they are shipped to hospitals so that contaminants cannot enter the packages. The substance that seals the package is referred to as the sealing die. Generally, sealing dies are heated with an electric heater. For the sealing process to be a success, the sealing die is maintained at an established temperature and must contact the package with a predetermined pressure for an established period of time. The period of time during which the sealing die contacts the package is called the dwell time. Assume that the ranges of parameters for an acceptable seal are the following:

Temperature: 150–170°C
Pressure: 60–70 psi
Dwell Time: 2.0–2.5 s

12. Assume that a file named **suture.dat** contains information on batches of sutures that have been rejected during a one-week period. Each line in the data file contains the batch number, the temperature, the pressure, and the dwell time for a rejected batch. A quality-control engineer would like to analyze this information to determine

    - the percent of the batches rejected due to temperature,
    - the percent rejected due to pressure, and
    - the percent rejected due to dwell time.

    If a specific batch is rejected for more than one reason, it should be counted in all applicable totals. Give the MATLAB statements to compute and print these three percentages. Use the following data to create **suture.dat**:

Batch Number	Temperature	Pressure	Dwell Time
24551	145.5°F	62.3	2.23
24582	153.7°F	63.2	2.52
26553	160.3°F	58.9	2.51
26623	159.5°F	58.9	2.01
26642	160.3°F	61.2	1.98

13. Modify the solution developed in Problem 12 so that it also prints the number of batches in each rejection category and the total number of batches rejected. (Remember that a rejected batch should appear only once in the total, but could appear in more than one rejection category.)
14. Confirm that the data in **suture.dat** relates only to batches that should have been rejected. If any batch should not be in the data file, print an appropriate message with the batch information.

TIMBER REGROWTH

A problem in timber management is to determine how much of an area to leave uncut so that the harvested area is reforested in a certain period of time. It is assumed that reforestation takes place at a known rate per year, depending on climate and soil conditions. A reforestation equation expresses this growth as a function of the amount of timber standing and the reforestation rate. For example, if 100 acres are left standing after harvesting and the reforestation rate is 0.05, then 105 acres are forested at the end of the first year. At the end of the second year, the number of acres forested is 110.25 acres. If $year_0$ is the acreage forested, then

$$
\begin{aligned}
year_1 &= year_0 + rate*year_0 = year_0*(1+rate) \\
year_2 &= year_1 + rate*year_1 = year_1*(1+rate) \\
&= year_0*(1+rate)*(1+rate) = year_0*(1+rate)^2 \\
year_3 &= year_2 + rate*year_2 = year_2*(1+rate) \\
&= year_0*(1+rate)^3 \\
year_n &= year_0*(1+rate)^n
\end{aligned}
$$

15. Assume that there are 14,000 acres total, with 2500 uncut acres and that the reforestation rate is 0.02. Print a table showing the number of acres reforested at the end of each year for a total of 20 years. You should also present your results in a bar graph, labeled appropriately.
16. Modify the program developed in Problem 15 that the user can enter the number of years to be used for the table.
17. Modify the program developed in Problem 15 so that the user can enter a number of acres, and the program will determine how many years are required for the number of acres to be forested. (You will need a loop for this one.)

SENSOR DATA

18. Suppose that a file named **`sensor.dat`** contains information collected from a set of sensors. Each row contains a set of sensor readings, with the first row containing values collected at 0 seconds, the second row containing values collected at 1.0 seconds, and so on. Write a program to print the subscripts of sensor data values with an absolute value greater than 20.0, using the **`find`** command.

POWER PLANT OUTPUT

The power output in megawatts from a power plant over a period of 8 weeks has been stored in a data file named **`plant.dat`**. Each line in the file represents data for one week and contains the output for day 1, day 2, through day 7.

19. Write a program that uses the power-plant output data and prints a report that lists the number of days with greater-than-average power output. The report should give the week number and the day number for each of these days, in addition to printing the average power output for the plant during the 8-week period.
20. Write a program that uses the power-plant output data and prints the day and week during which the maximum and minimum power output occurred. If the maximum or minimum power output occurred on more than one day, the program should print all the days involved.
21. Write a program that uses the power-plant output data to print the average power output for each week. Also print the average power output for day 1, day 2, and so on.

CHAPTER

6 Matrix Computations

Objectives

After reading this chapter, you should be able to

- use MATLAB functions to generate special matrices,
- perform operations that apply to an entire matrix as a unit, and
- solve simultaneous equations using MATLAB matrix operations.

ENGINEERING ACHIEVEMENT: MANNED SPACE FLIGHT

Some of the greatest achievements of engineering over the last few decades have included manned space flight. These achievements began with the first manned space flight that occurred on April 12, 1961, when the Russians launched a spacecraft manned by Yuri Gagarin that orbited the Earth. The first American manned space flight occurred on May 5, 1961, when Alan Shepard completed a suborbital flight. On February 20, 1962, John Glenn became the first American to orbit the Earth. The first Moon landing occurred on July 20, 1969. The Moon landing was probably the most complex and ambitious engineering project ever attempted. Major breakthroughs were required in the design of the Apollo spacecraft, the lunar lander, and the three-stage Saturn V rocket. Even the design of the space suit was a major engineering project, resulting in a system that included a three-piece space suit and backpack, which together weighed 190 pounds. In this chapter, we include problems to show how matrices are useful in analyzing the weight of spacecraft components and in solving equations needed for the design of electrical circuits used in spacecraft sensors.

6.1 SPECIAL MATRICES

In this chapter, we present matrix operations that use the entire matrix, such as matrix multiplication. (In Chapter 2, we presented element-by-element operations that used only one element from the matrix at a time.) We will show how these matrix operations can be used to solve a number of different types of problems, from computing dot products and matrix products, to solving systems of linear equations. However, before we do that, we present a group of MATLAB functions that generate special matrices. Some of these functions will be used in later sections of this chapter.

6.1.1 Matrices of Zeros and Ones

We often need to create a matrix that is filled with zeros, or that is filled with ones. MATLAB has two functions to make that an easy task to perform. The **zeros** function generates a matrix containing all zeros. If the argument to the function is a scalar, as in **zeros(6)**, the function will generate a square matrix using the argument as both the number of rows and the number of columns. If the function has two scalar arguments, as in **zeros(m,n)**, the function will generate a matrix with **m** rows and **n** columns. Because the **size** function returns two scalar arguments that represent the number of rows and columns in a matrix, we can use the **size** function to generate a matrix of zeros that is the same size as another matrix. The following MATLAB statements and the corresponding values displayed illustrate these various cases:

```
A = zeros(3)
A =
    0  0  0
    0  0  0
    0  0  0
B = zeros(3,2)
B =
    0  0
    0  0
    0  0
C = [1,2,3; 4,2,5]
C =
    1  2  3
    4  2  5
D = zeros(size(C))
D =
    0  0  0
    0  0  0
```

The **ones** function generates a matrix containing all ones, just as the **zeros** function generates a matrix containing all zeros. If the argument to the function is a scalar, as in **ones(6)**, the function will generate a square matrix using the argument as both the number of rows and the number of columns. If the function has two scalar arguments, as in **ones(m,n)**, the function will generate a matrix with **m** rows and **n** columns. To generate a matrix of ones that is the same size as another matrix, use the **size** function to determine the correct number of rows and columns. The following MATLAB statements illustrate these various cases:

```
A = ones(3)
A =
    1  1  1
    1  1  1
    1  1  1
```

```
B = ones(3,2)
B =
   1  1
   1  1
   1  1
C = [1,2,3; 4,2,5]
C =
   1  2  3
   4  2  5
D = ones(size(C))
D =
   1  1  1
   1  1  1
```

6.1.2 Identity Matrix

An **identity matrix** is a matrix with ones on the main diagonal and zeros everywhere else. For example, the following matrix is an identity matrix with four rows and four columns:

$$\begin{bmatrix} 1 & 0 & 0 & 0 \\ 0 & 1 & 0 & 0 \\ 0 & 0 & 1 & 0 \\ 0 & 0 & 0 & 1 \end{bmatrix}$$

Note that the **main diagonal** is the diagonal containing elements in which the row number is the same as the column number. Therefore, the subscripts for elements on the main diagonal in this example are (1,1), (2,2), (3,3), and (4,4).

In MATLAB, identity matrices can be generated using the **eye** function. The arguments of the **eye** function are similar to those for the **zeros** and the **ones** functions. If the argument to the function is a scalar, as in **eye(6)**, the function will generate a square matrix using the argument as both the number of rows and the number of columns. If the function has two scalar arguments, as in **eye(m,n)**, the function will generate a matrix with **m** rows and **n** columns. To generate an identity matrix that is the same size as another matrix, use the **size** function to determine the correct number of rows and columns. Although most applications use a square identity matrix, the definition can be extended to nonsquare matrices. The following statements illustrate these various cases:

```
A = eye(3)
A =
   1  0  0
   0  1  0
   0  0  1
B = eye(3,2)
B =
   1  0
   0  1
   0  0
C = [1,2,3; 4,2,5]
C =
   1 2 3
   4 2 5
```

```
D = eye(size(C))
D =
   1 0 0
   0 1 0
```

HINT

> We recommend that you do not name an identity matrix **`i`**, because **`i`** will no longer represent $\sqrt{-1}$ in any statements that follow.

6.1.3 Diagonal Matrices

The **`diag`** function can be used to extract one of the diagonals of a matrix, or to create a **diagonal matrix**. To extract the main diagonal from a matrix, consider the following statements and values displayed:

```
A = [1, 0, 5; 7, 4, -2; 3, -1, 1];
diag(A)
ans =
     1
     4
     1
```

Other diagonals can be specified by passing a second parameter **`k`** to the function that denotes the position of the diagonal from the main diagonal (**`k`** = 0). We illustrate this using the example matrix **`A`**:

```
diag(A,1)
ans =
     0
    -2
```

If the first argument to **`diag`** is a vector **`V`**, then this function generates a square matrix. If the second parameter **`k`** is equal to zero, then the elements of **`V`** are placed on the main diagonal, and if **`k`** > 0, they are placed above the main diagonal. If **`k`** < 0, they are placed below the main diagonal. Thus, we have

```
B = [1, 2, 3]
B =
   1 2 3
diag(B)
ans =
     1 0 0
     0 2 0
     0 0 3
diag(B,1)
ans =
     0 1 0 0
     0 0 2 0
     0 0 0 3
     0 0 0 0
```

There are two additional functions that are often useful when extracting diagonals from a matrix. The **fliplr** function flips a matrix from left to right; the **flipud** function flips a matrix from up to down. For example, these statements display the matrix **C**, and the matrix generated by these functions:

```
C = [1, 2, 3; 4, 2, 5]
C =
   1 2 3
   4 2 5
D = fliplr(C)
D =
   3 2 1
   5 2 4
E = flipud(C)
E =
   4 2 5
   1 2 3
```

We will illustrate the use of one of these functions in the next section.

6.1.4 Magic Squares

MATLAB includes a matrix function called **magic** that generates a **magic square**—one in which the sum of all of the columns is the same, as is the sum of all of the rows. For example, consider the following statements that generate a magic square with four rows and four columns, and then computes the sums of the columns:

```
A = magic(4)
A =
   16   2   3  13
    5  11  10   8
    9   7   6  12
    4  14  15   1
sum(A)
ans =
     34 34 34 34
```

To find the sum of the rows, we need to transpose the matrix:

```
sum(A')
ans =
     34 34 34 34
```

Not only is the sum of all of the columns and rows the same; but also the sum of the left-to-right main diagonal is the same as the sum of the right-to-left main diagonal. The diagonal from left to right is

```
diag(A)
ans =
     16
     11
      6
      1
```

Finding the sum of the diagonal reveals the same number as the sum of the rows and columns:

```
sum(diag(A))
ans =
     34
```

Finally, to find the diagonal from lower left to upper right, we first have to flip the matrix and then find the sum of the diagonal:

```
fliplr(A)
ans =
     13    3    2   16
      8   10   11    5
     12    6    7    9
      1   15   14    4
diag(ans)
ans =
     13
     10
      7
      4
sum(ans)
ans =
     34
```

One of the earliest documented examples of a magic square is in a woodcut (shown in Figure 6.1) by Albrecht Dürer, created in 1514. Scholars believe the square was a reference to alchemical concepts popular at the time. The date of the woodcut is included in the two middle squares of the bottom row. Magic squares have fascinated both professional and amateur mathematicians for centuries. For example, Benjamin Franklin experimented with magic squares. You can create magic squares

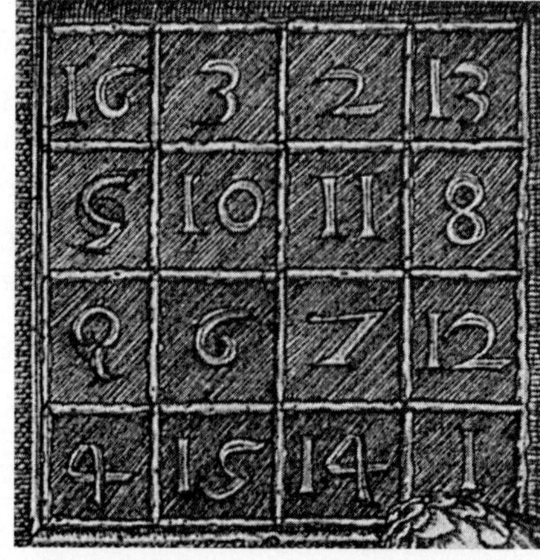

Figure 6.1
Melancholia, by Albrect Dürer, 1514, and a close-up of an embedded magic square.

of any size greater than 2 × 2, using MATLAB. However, other magic squares are possible–MATLAB's solution is not the only one.

6.2 MATRIX OPERATIONS AND FUNCTIONS

Many engineering computations use a matrix as a convenient way to represent a set of data and to perform operations on that data. In this chapter, we are primarily interested in matrices that are not vectors; that is, they have more than one row and more than one column. In Chapter 2, we presented scalar operations that are performed element-by-element. In this section, we present matrix operations and matrix multiplication.

6.2.1 Transpose

The **transpose** of a matrix is a new matrix in which the rows of the original matrix are the columns of the new matrix. Mathematically, we use a superscript **T** after the name of a matrix to refer to the transpose of the matrix. For example, consider the following matrix and its transpose:

$$\mathbf{A} = \begin{bmatrix} 2 & 5 & 1 \\ 7 & 3 & 8 \\ 4 & 5 & 21 \\ 16 & 13 & 0 \end{bmatrix} \qquad \mathbf{A}^{\mathbf{T}} = \begin{bmatrix} 2 & 7 & 4 & 16 \\ 5 & 3 & 5 & 13 \\ 1 & 8 & 21 & 0 \end{bmatrix}$$

If we consider a couple of the elements, we see that the value in position (3,1) of **A** has now moved to position (1,3) of $\mathbf{A}^{\mathbf{T}}$, and the value in position (4,2) of **A** has now moved to position (2,4) of $\mathbf{A}^{\mathbf{T}}$. In general terms, the row and column subscripts are interchanged to form the transpose; hence, the value in position (i, j) is moved to position (j, i).

In MATLAB, the transpose of the matrix `A` is denoted by `A'`. Observe that the transpose will have a different size than the original matrix if the original matrix is not a square matrix. We frequently use the transpose operation to convert a row vector to a column vector or a column vector to a row vector.

6.2.2 Dot Product

The **dot product** is a scalar computed from two vectors of the same size. This scalar is the sum of the products of the values in corresponding positions in the vectors, as shown in the following summation equation, which assumes that there are n elements in the vectors **A** and **B**:

$$\text{dot product} = \mathbf{A} \cdot \mathbf{B} = \sum_{i=1}^{n} a_i b_i$$

In MATLAB, we can compute the dot product with the following statement:

```
dot_product = sum(A.*B);
```

Recall that `A.*B` contains the results of an element-by-element multiplication of `A` and `B`. When `A` and `B` are both row vectors or are both column vectors, `A.*B` is also a vector. We then sum the elements in this vector, thus yielding the dot product. The `dot` function may also be used to compute the dot product:

```
dot(A,B);
```

To illustrate, assume that **A** and **B** are the following vectors:

$$\mathbf{A} = [4 \;\; -1 \;\; 3] \qquad \mathbf{B} = [-2 \;\; 5 \;\; 2]$$

The dot product is then

$$\begin{aligned}\mathbf{A}\cdot\mathbf{B} &= 4\cdot(-2) + (-1)\cdot 5 + 3\cdot 2\\ &= (-8) + (-5) + 6\\ &= -7\end{aligned}$$

You can test this result by defining the two vectors, and then typing

```
dot(A,B)
```

EXAMPLE 6.1

CALCULATING MASS OF A SPACECRAFT

The mass of space vehicles, such as the one shown in Figure 6.2, is extremely important. Entire groups of people in the design process keep track of the location and mass of every nut and bolt. This information is used to determine the center of gravity of the vehicle in addition to its total mass. One reason the center of gravity is important is that rockets tumble if the center of gravity is behind the center of pressure. You can demonstrate this with a paper airplane. Put a paperclip on the tail of the paper airplane and observe how the flight pattern changes.

Figure 6.2
A Titan satellite launch vehicle.

Although finding the center of gravity is a fairly straightforward calculation, it becomes more complex when you realize that the mass of the vehicle and the distribution of mass changes as the fuel is burned.

In this example, we will only find the total mass of some of the components used in a complex space vehicle, as shown below:

Item	Amount	Mass (g)
Bolt	3	3.50
Screw	5	1.50
Nut	2	0.79
Bracket	1	1.75

The total mass is really a dot product. You need to multiply each amount times the corresponding mass, and then add them up. Write a MATLAB program to find the mass of this list of components, using matrix math.

SOLUTION

1. Problem Statement

Find the total mass for a set of specified components.

2. Input/Output Description

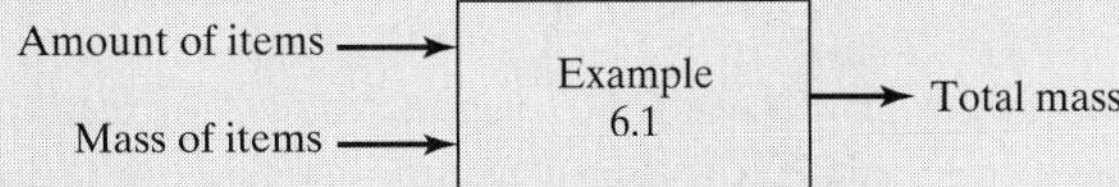

3. Hand Example

Here is a table of items, how many of these items are being used, and the mass of each individual item:

Item	Amount		Mass (g)	Totals (g)
Bolt	3	×	3.50	= 10.50
Screw	5	×	1.50	= 7.50
Nut	2	×	0.79	= 1.58
Bracket	1	×	1.75	= 1.75
				21.33

At the end of each row, we compute the mass for that item, and then add these up at the end of the table.

4. MATLAB Solution

We use an M-file for our solution so that it will be easy to store, and easy to modify for additional data sets. We demonstrate two ways to compute the total mass—using vector computations and using the dot product function.

```
%-----------------------------------------------------------------
% Example 6_1 - This program determines the mass
% for a set of items given the quantity and mass
```

(continued)

```
% of each individual item.
%
clear, clc
%
% Define the item and mass vectors.
amount = [3, 5, 2, 1];
mass = [3.5, 1.5, 0.79, 1.75];
%
% Compute the mass using vector computations.
total_mass_1 = sum(amount.*mass)
%
% Compute the mass with the dot product function.
total_mass_2 = dot(amount,mass)
%-----------------------------------------------------------
```

5. Testing

The output for this program is:

```
total_mass_1 =
   21.3300
total_mass_2 =
   21.3300
```

Both approaches give the same result and agree with the hand example. Now that we know the program works, we can use it for any number of items.

6.2.3 Matrix Multiplication

Matrix multiplication is not accomplished by multiplying corresponding elements of the matrices. In matrix multiplication, the value in position $c(i,j)$ of the product **C** of two matrices **A** and **B** is the dot product of row i of the first matrix and column j of the second matrix, as shown in the following summation equation:

$$c_{i,j} = \sum_{k=1}^{N} a_{ik} b_{kj}$$

Because the dot product requires that the vectors have the same number of elements, the first matrix **A** must have the same number of elements N in each row as there are in each column of the second matrix **B**. Thus, if **A** and **B** both have five rows and five columns, their product has five rows and five columns respectively. Furthermore, for these matrices, we can compute both **AB** and **BA**, but in general they will not be equal.

If **A** has two rows and three columns and **B** has three rows and three columns, the product **AB** will have two rows and three columns. To illustrate, consider the following matrices:

$$\mathbf{A} = \begin{bmatrix} 2 & 5 & 1 \\ 0 & 3 & -1 \end{bmatrix} \qquad \mathbf{B} = \begin{bmatrix} 1 & 0 & 2 \\ -1 & 4 & -2 \\ 5 & 2 & 1 \end{bmatrix}$$

The first element in the product **C** = **AB** is

$$\begin{aligned} c_{1,1} &= \sum_{k=1}^{3} a_{1k} b_{k1} \\ &= a_{1,1}\, b_{1,1} + a_{1,2}\, b_{2,1} + a_{1,3}\, b_{3,1} \\ &= 2\cdot 1 + 5\cdot(-1) + 1\cdot 5 \\ &= 2 \end{aligned}$$

Similarly, we can compute the rest of the elements in the product of **A** and **B**:

$$\mathbf{AB} = \mathbf{C} = \begin{bmatrix} 2 & 22 & -5 \\ -8 & 10 & -7 \end{bmatrix}$$

In this example, we cannot compute **BA**, because **B** does not have the same number of elements in each row as **A** has in each column.

An easy way to decide whether a matrix product exists is to write the sizes of the two matrices side by side. Then, if the two inside numbers are the same, the product exists, and the size of the product is determined by the two outside numbers. To illustrate, in the previous example, the size of **A** is 2×3, and the size of **B** is 3×3. Therefore, if we want to compute **AB**, we write the sizes side by side:

$$2 \times 3, 3 \times 3$$

The two inner numbers are both the value 3, so **AB** exists, and its size is determined by the two outer numbers, 2×3. If we want to compute **BA**, we again write the sizes side by side:

$$3 \times 3, 2 \times 3$$

The two inner numbers are not the same, so **BA** does not exist. If the two inner numbers are the same, then **A** is said to be **conformable** for multiplication to **B**.

In MATLAB, matrix multiplication is denoted by an asterisk. Thus, the command to perform matrix multiplication of matrices **A** and **B** is

```
A*B;
```

For example, to generate the matrices in our previous example, and then compute the matrix product, we can use the following MATLAB statements:

```
A = [2, 5, 1; 0, 3, -1];
B = [1, 0, 2; -1, 4, -2; 5, 2, 1];
C = A*B
C =
    2 22 -5
   -8 10 -7
```

Note that `B*A` does not exist, because the number of columns of `B` does not equal the number of rows of `A`. In other words, `B` is not conformable for multiplication with `A`. Execute this MATLAB command:

```
C = B*A;
```

You will get the following warning message:

```
??? Error using ==> *
Inner matrix dimensions must agree.
```

In general, the matrix product **AB** is not equal to the matrix product **BA**. However, the products **AB** and **BA** are equal if one of the matrices is an identity matrix. We demonstrate this with the following MATLAB statements and their corresponding values:

```
A = [1, 0, 2; -1, 4, -2; 5, 2, 1];
I = eye(3)
I =
   1 0 0
   0 1 0
   0 0 1
A*I
ans =
        1 0  2
       -1 4 -2
        5 2  1
I*A
ans =
        1 0  2
       -1 4 -2
        5 2  1
```

EXAMPLE 6.2

COMPUTING MASS FOR SEVERAL VENDORS

Suppose you would like to know which commercial vendor offers the least overall mass total for the items you'll use. Your list of items stays the same, but the mass of each item is different because they are purchased from different vendors. An example is shown below:

Item	Amount	Mass (g) Vendor A	Mass (g) Vendor B	Mass (g) Vendor C
Bolt	3	3.50	2.98	2.50
Screw	5	1.50	1.75	1.60
Nut	2	0.79	1.25	0.99
Bracket	1	1.75	0.95	1.25

The total mass for each vendor is a dot product. You need to multiply each amount times the corresponding mass, and then add them up. But it would be nice to do just one calculation. Matrix multiplication is the answer. We will need to define the amount vector as a row, but the mass matrix will be a 4 $\times$ 3 matrix.

SOLUTION

1. Problem Statement

Find the total mass for each vendor.

2. Input/Output Description

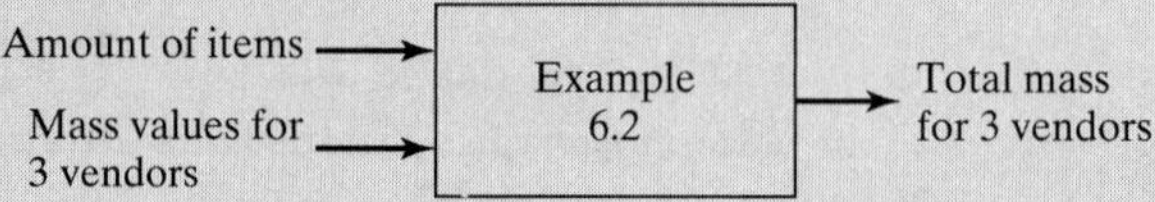

3. Hand Example

In the table below we compute the mass for vendor A:

Item	Amount		Mass (g)	Totals (g)
Bolt	3	×	3.50	=10.50
Screw	5	×	1.50	= 7.50
Nut	2	×	0.79	= 1.58
Bracket	1	×	1.75	= 1.75
				21.33

Similar computations give the mass for other two vendors:

$$\text{mass for vendor B} = 21.14 \text{ g}$$
$$\text{mass for vendor C} = 18.73 \text{ g}$$

4. MATLAB Solution

We use an M-file for our solution. As we convert our hand solution into a MATLAB solution, we need to think about the matrix sizes. The item matrix will have one row and four columns; the vendor matrix will have four rows and three columns. If we check the size of the matrix multiplication result, we have $(1 \times 4)\,(4 \times 3)$, which gives a (1×3) solution which contains the total mass for each of the three vendors.

```
%-----------------------------------------------------------
% Example 6_2 - This program determines the mass
% for a set of items given the quantity and mass
% of each individual item, for a group of vendors.
%
clear, clc
%
% Define the item and mass vectors
amount = [3, 5, 2, 1];
vendor_1 = [3.50, 1.50, 0.79, 1.75]';
vendor_2 = [2.98, 1.75, 1.25, 0.95]';
vendor_3 = [2.50, 1.60, 0.99, 1.25]';
vendor_matrix = [vendor_1, vendor_2, vendor_3];
%
% Compute the values for the three vendors.
vendor_mass = amount*vendor_matrix;
disp('Vendor Mass Amounts')
disp(vendor_mass)
%-----------------------------------------------------------
```

5. Testing

The output for the this program is shown below

```
Vendor Mass Amounts
21.3300 21.1400 18.7300
```

Based on this solution, vendor C is the best choice in terms of mass. (The best choice overall may not be vendor C because of other factors, including quality and cost.)

EXAMPLE 6.3

COMPARING MASS FOR COMPETING DESIGNS

Assume that two engineers are promoting competing designs, as shown below:

Item	Amount Design A	Amount Design B	Mass (g) Vendor A	Mass (g) Vendor B	Mass (g) Vendor C
Bolt	3	5	3.50	2.98	2.50
Screw	5	2	1.50	1.75	1.60
Nut	2	3	0.79	1.25	0.99
Bracket	1	2	1.75	0.95	1.25

Now, for each design, we need to multiply each amount times the corresponding mass. Again, matrix multiplication is a good choice for computing the answers.

SOLUTION

1. Problem Statement

Find the total mass for each vendor.

2. Input/Output Description

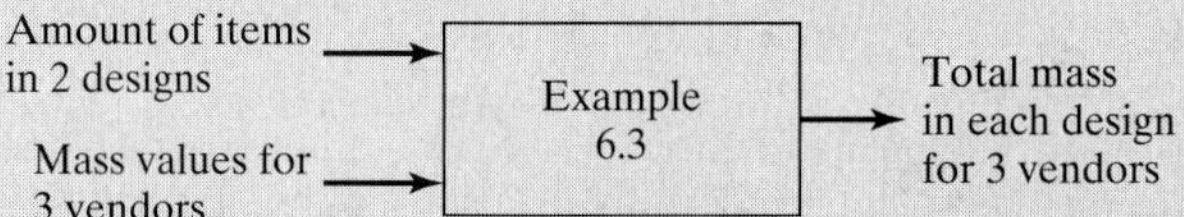

3. Hand Example

In the previous problem, we computed the mass for the first design for the three vendors. In the table below, we compute the mass for the second design, for vendor A.

Item	Amount		Mass (g)	Totals (g)
Bolt	5	×	3.50	= 17.50
Screw	2	×	1.50	= 3.00
Nut	3	×	0.79	= 2.37
Bracket	2	×	1.75	= 3.50
				26.37

Similar computations give the mass for other two vendors, for the second design:

$$\text{mass for vendor B} = 24.0500 \text{ g}$$
$$\text{mass for vendor C} = 21.1700 \text{ g}$$

4. MATLAB Solution

We use an M-file for our solution. As we convert our hand solution into a MATLAB solution, we need to think about the matrix sizes. The "amount matrix" will have two rows (one for each design) and four columns (to represent the number of items in each design). The "vendor matrix" will have four rows (one for the mass of each of the four items) and three columns (one for each vendor). If we check the size of the matrix multiplication result, we have $(2 \times 4)(4 \times 3)$, which gives a (2×3) solution which contains the mass of the three vendors, for each design.

```
%-------------------------------------------------------------
% Example 6_3 - This program determines the mass
% for competing designs. It does this by specifying
% the amounts of each item for each design, and
% then specifying the mass for each item
% for a group of vendors.
%
clear, clc
%
% Define the item amounts and mass vectors
design_1_items = [3, 5, 2, 1];
design_2_items = [5, 2, 3, 2];
designs = [design_1_items; design_2_items];
vendor_1 = [3.50, 1.50, 0.79, 1.75}';
vendor_2 = [2.98, 1.75, 1.25, 0.95]';
vendor_3 = [2.50, 1.60, 0.99, 1.25]';
vendor_matrix = [vendor_1, vendor_2, vendor_3];
%
% Compute the design values for the three vendors.
mass = designs*vendor_matrix;
%
design_1 = mass(1,:)
design_2 = mass(2,:)
%-------------------------------------------------------------
```

5. Testing

The output for this program is shown below

```
design_1
   21.3300   21.1400   18.7300
design_2
   26.3700   24.0500   21.1700
```

It is important in this program to be sure to identify which numbers go with each design, and thus we printed these separately. If we had just printed the matrix of values, it would not necessarily be clear which ones went with each design.

6.2.4 Matrix Powers

Recall that if **A** is a matrix, then the expression **A.^2** squares each element in **A**. If we want to square the matrix—that is, compute **A*A**—we can use the operation **A^2**. Thus, **A^4** is equivalent to **A*A*A*A.** To perform a matrix multiplication between two matrices, the number of rows in the first matrix must be the same value as the number of columns in the second matrix. Therefore, to raise a matrix to a power, the number of rows must equal the number of columns, and thus the matrix must be a square matrix. For example, consider the following statements and corresponding display:

```
A = [1 2; 3, 4];
C = A^2
```

```
C =
     7   10
    15   22
D = A.^2;
D =
     1    4
     9   16
```

Thus, it is always important when multiplying matrices to be sure that we distinguish between element-by-element multiplication and actual matrix multiplication.

6.2.5 Matrix Inverse

By definition, the **inverse** of a square matrix **A** is the matrix $\mathbf{A}^{-1}$ such that the matrix products $\mathbf{AA}^{-1}$ and $\mathbf{A}^{-1}\mathbf{A}$ are both equal to the identity matrix. For example, consider the following two matrices **A** and **B**:

$$\mathbf{A} = \begin{bmatrix} 2 & 1 \\ 4 & 3 \end{bmatrix} \qquad \mathbf{B} = \begin{bmatrix} 1.5 & -0.5 \\ -2 & 1 \end{bmatrix}$$

If we compute the products **AB** and **BA**, we obtain the following matrices (do the matrix multiplications by hand to be sure you follow the steps):

$$\mathbf{AB} = \begin{bmatrix} 1 & 0 \\ 0 & 1 \end{bmatrix} \qquad \mathbf{BA} = \begin{bmatrix} 1 & 0 \\ 0 & 1 \end{bmatrix}$$

Therefore, **A** and **B** are inverses of each other, or $\mathbf{A} = \mathbf{B}^{-1}$ and $\mathbf{B} = \mathbf{A}^{-1}$.

Computing the inverse of a matrix is a tedious process; fortunately, MATLAB contains an **`inv`** function that performs the computations for us. (We do not present the steps for computing an inverse in this text. Refer to a linear algebra text if you are interested in the techniques for computing an inverse.) Thus, if we execute **`inv(A)`** using the matrix **A** defined previously, the result will be another matrix containing the inverse. If we then compute the inverse of that matrix, the result should be the original matrix **A**. Also, recall that the product of a matrix with its inverse equals the identity matrix. We can illustrate these relationships with the following statements:

```
A = [1, 0, 2; -1, 4, -2; 5, 2, 1]
A =
     1   0   2
    -1   4  -2
     5   2   1
B = inv(A)
B =
    -0.2222   -0.1111    0.2222
     0.2500    0.2500    0.0000
     0.6111    0.0556   -0.1111
A*B
ans =
      1.0000         0    0.0000
     -0.0000    1.0000    0.0000
     -0.0000   -0.0000    1.0000
```

```
B*A
ans =
      1.0000   0.0000    0.0000
      0.0000   1.0000   -0.0000
     -0.0000        0    1.0000
```

There are matrices for which an inverse does not exist; these matrices are called **singular** or **ill-conditioned matrices**. When you attempt to compute the inverse of an ill-conditioned matrix in MATLAB, an error message is printed.

6.2.6 Determinants

A **determinant** is a scalar value computed from the entries in a square matrix. Determinants have various applications in engineering, including computing inverses and solving systems of simultaneous equations. For a 2 × 2 matrix **A**, the determinant is defined to be

$$|\mathbf{A}| = a_{1,1}a_{2,2} - a_{2,1}a_{1,2}$$

Therefore, the determinant of **A**, or $|\mathbf{A}|$, is equal to 8 for the following matrix:

$$\mathbf{A} = \begin{bmatrix} 1 & 3 \\ -1 & 5 \end{bmatrix}$$

For a 3 × 3 matrix **A**, the determinant is defined to be

$$\begin{aligned} |\mathbf{A}| = \; & a_{1,1}a_{2,2}a_{3,3} + a_{1,2}a_{2,3}a_{3,1} + a_{1,3}a_{2,1}a_{3,2} \\ & - a_{3,1}a_{2,2}a_{1,3} - a_{3,2}a_{2,3}a_{1,1} - a_{3,3}a_{2,1}a_{1,2} \end{aligned}$$

If

$$\mathbf{A} = \begin{bmatrix} 1 & 3 & 0 \\ -1 & 5 & 2 \\ 1 & 2 & 1 \end{bmatrix}$$

then $|\mathbf{A}|$ is equal to $5 + 6 + 0 - 0 - 4 - (-3)$, or 10.

A more involved process is necessary for computing determinants of matrices with more than three rows and columns. We do not include a discussion of the process for computing a general determinant here, because MATLAB will automatically compute a determinant using the **`det`** function, with a square matrix as its argument, as in **`det(A)`**.

6.3 SOLUTIONS TO SYSTEMS OF LINEAR EQUATIONS

Consider the following system of three equations with three unknowns:

$$\begin{aligned} 3x \; +2y \; -z &= 10 \\ -x \; +3y \; +2z &= 5 \\ x \; -y \; -z &= -1 \end{aligned}$$

We can rewrite this system of equations using the following matrices:

$$\mathbf{A} = \begin{bmatrix} 3 & 2 & -1 \\ -1 & 3 & 2 \\ 1 & -1 & -1 \end{bmatrix} \quad \mathbf{X} = \begin{bmatrix} x \\ y \\ z \end{bmatrix} \quad \mathbf{B} = \begin{bmatrix} 10 \\ 5 \\ -1 \end{bmatrix}$$

Using matrix multiplication, the **system of equations** can then be written as **AX** = **B**. Go through the multiplication to convince yourself that this matrix equation yields the original set of equations.

To simplify the notation, we designate the variables as x_1, x_2, x_3, and so on. Rewriting the initial set of equations using this notation, we have

$$\begin{aligned} 3x_1 \quad +2x_2 \quad -x_3 &= 10 \\ -x_1 \quad +3x_2 \quad +2x_3 &= 5 \\ x_1 \quad -x_2 \quad -x_3 &= -1 \end{aligned}$$

This set of equations is then represented by the matrix equation **AX** = **B**, where **X** is the column vector $[x_1, x_2, x_3]^T$. We now present two methods for solving a system of N equations with N unknowns.

6.3.1 Solution Using the Matrix Inverse

One way to solve a system of equations is by using the matrix inverse. For example, assume that **A**, **X**, and **B** are the matrices defined earlier in this section:

$$\mathbf{A} = \begin{bmatrix} 3 & 2 & -1 \\ -1 & 3 & 2 \\ 1 & -1 & -1 \end{bmatrix} \qquad \mathbf{X} = \begin{bmatrix} x_1 \\ x_2 \\ x_3 \end{bmatrix} \qquad \mathbf{B} = \begin{bmatrix} 10 \\ 5 \\ -1 \end{bmatrix}$$

Then **AX** = **B**. If we premultiply both sides of this matrix equation by $\mathbf{A}^{-1}$, we have $\mathbf{A}^{-1}\mathbf{AX} = \mathbf{A}^{-1}\mathbf{B}$. However, because $\mathbf{A}^{-1}\mathbf{A}$ is equal to the identity matrix **I**, we have $\mathbf{IX} = \mathbf{A}^{-1}\mathbf{B}$, or $X = A^{-1}B$. In MATLAB, we can compute this solution with the following command:

```
X = inv(A)*B;
```

As an example, we will solve the following system of equations:

$$\begin{aligned} 3x_1 + 5x_2 &= -7 \\ 2x_1 - 4x_2 &= 10 \end{aligned}$$

Type the following MATLAB commands to define `A` and `B`, and then solve the system of equations:

```
A = [3, 5; 2, -4];
B = [-7, 10]';
X = inv(A)*B
X =
      1.0000
     -2.0000
```

Note that `B` is defined to be a column matrix. Substitute these values for `X` back into the original equations to convince yourself that these values represent the solution to the system of equations.

EXAMPLE 6.4

SOLVING ELECTRIC CIRCUIT EQUATIONS

Solving an electrical circuit problem can often result in a set of simultaneous equations to solve. For example, consider the electrical circuit in Figure 6.3. It contains a single voltage source and five resistors. You can analyze this circuit by dividing it up into smaller pieces and applying two basic electrical facts:

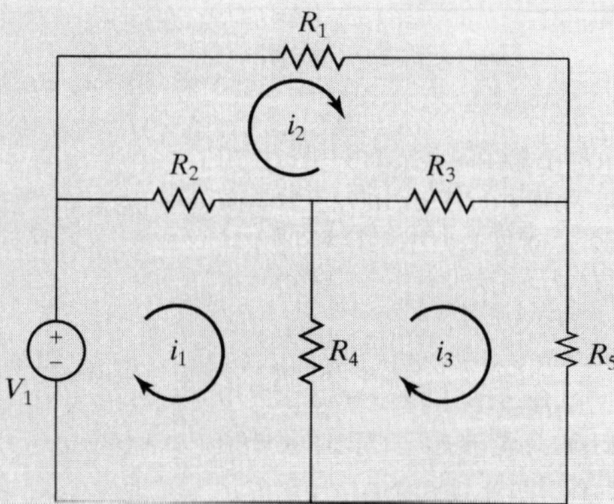

Figure 6.3
An electrical circuit.

$$\Sigma \textit{ voltage} \text{ around a circuit must be zero}$$

$$\text{Voltage} = \text{current} \times \text{resistence}, \; V = iR$$

Following the lower left-hand loop results in our first equation:

$$-V_1 + R_2(i_1 - i_2) + R_4(i_1 - i_3) = 0$$

Following the upper loop results in our second equation:

$$R_1 i_2 + R_3(i_2 - i_3) + R_2(i_2 - i_1) = 0$$

Finally, following the lower right-hand loop results in the last equation:

$$R_3(i_3 - i_2) + R_5 i_3 + R_4(i_3 - i_1) = 0$$

Since we know all the resistances (R values) and the voltage, we have three equations and three unknown currents. Now we need to rearrange the equations so that they are in a form where we can perform a matrix solution—in other words, we need to isolate the current variables:

$$(R_2 + R_4)i_1 + (-R_2)i_2 + (-R_4)i_3 = V_1$$

$$(-R_2)i_1 + (R_1 + R_2 + R_3)i_2 + (-R_3)i_3 = 0$$

$$(-R_4)i_1 + (-R_3)i_2 + (R_3 + R_4 + R_5)i_3 = 0$$

Write a MATLAB program to solve these equations, using the matrix inverse method. Allow the user to enter the five values of R and the voltage from the keyboard.

SOLUTION

1. Problem Statement

Find the three currents for the circuit shown in Figure 6.3.

2. Input/Output Description

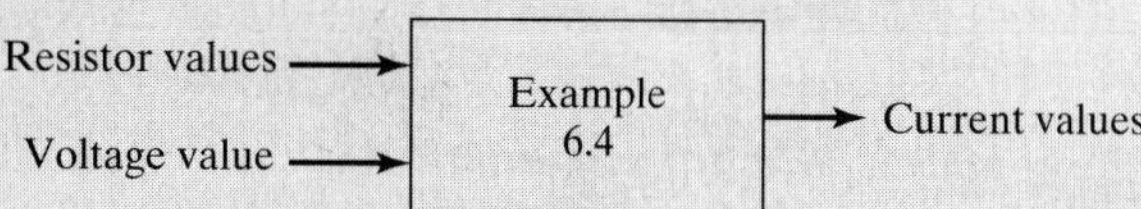

(*continued*)

3. Hand Example

For a hand example, assume that all the resistor values are 1 ohm, and that the voltage source is 5 volts. The set of equations then becomes the following:

$$
\begin{aligned}
2i_1 - i_2 - i_3 &= 5 \\
-i_1 + 3i_2 - i_3 &= 0 \\
-i_1 - i_2 + 3i_3 &= 0
\end{aligned}
$$

To solve this set of equations without MATLAB, you can either do it by hand, or use your calculator. In either case, you should get an answer of $i_1 = 5$, $i_2 = 2.5$, $i_3 = 2.5$. A negative value for current indicates that the current is flowing in a direction opposite to the one we chose as positive.

4. MATLAB Solution

We develop the solution as an M-file in order to be able to store it and to easily modify it for additional problem solutions.

```
%-----------------------------------------------------------
% Example 6_4 - This program reads resistor and voltage
% values and then computes the corresponding current
% values for a specified electrical circuit.
%
clear, clc
%
% Enter resistor values.
R1 = input('Input resistor value R1 in ohms:');
R2 = input('Input resistor value R2 in ohms:');
R3 = input('Input resistor value R3 in ohms:');
R4 = input('Input resistor value R4 in ohms:');
R5 = input('Input resistor value R5 in ohms:');
V = input('Input the value of voltage in volts:');
%
% Initialize the matrix A and vector B using AX = B form.
A = [(R2+R4), -R2, -R4;
     -R2, (R1+R2+R3), -R3;
     -R4, -R3, (R3+R4+R5)];
B = [V, 0, 0]';
current = inv(A)*B;
disp('Current:')
disp(current)
%-----------------------------------------------------------
```

5. Testing

If we test this program using the data from our hand example, the interaction will be the following:

```
Input resistor value R1 in ohms: 1
Input resistor value R2 in ohms: 1
Input resistor value R3 in ohms: 1
Input resistor value R4 in ohms: 1
```

```
Input resistor value R5 in ohms: 1
Input the value of voltage V in volts: 5
Current:
    5.0000
    2.5000
    2.5000
```

These values should match those you computed for the hand example. We now test the program with a different set of values:

```
Input resistor value R1 in ohms: 2
Input resistor value R2 in ohms: 4
Input resistor value R3 in ohms: 6
Input resistor value R4 in ohms: 8
Input resistor value R5 in ohms: 10
Input the value of voltage V in volts: 10
Current:
    1.6935
    0.9677
    0.8065
```

6.3.2 Solution Using Matrix Left Division

Another way to solve a system of linear equations is to use the **matrix left division** operator:

```
X = A\B;
```

This method produces the solution using Gaussian elimination, without forming the inverse. Using the matrix division operator is more efficient than using the matrix inverse and produces a greater numerical accuracy.

As an example, we will solve the same system of equations used in the previous example:

$$3x_1 + 5x_2 = -7$$

$$2x_1 - 4x_2 = 10$$

However, now solve for **X** by using matrix left division:

```
A = [3, 5; 2, -4];
B = [-7, 10]';
X = A\B
X =
    1
   -2
```

To confirm that the values of **X** do indeed solve each equation, we can multiply **A** by **X** using the expression **A*X**. The result is the column vector **[−7, 10]'**.

If there is not a unique solution to a system of equations, an error message is displayed. The solution vector may contain values of NaN, ∞ or $-\infty$, depending on the values of the matrices **A** and **B**.

SUMMARY

In this chapter, we presented matrix functions to create matrices of zeros, matrices of ones, identity matrices, diagonal matrices, and magic squares. We also defined the transpose, the inverse, and the determinant of a matrix, and presented functions to compute them. We also presented functions for flipping a matrix from left to right, and for flipping it from top to bottom. We defined the dot product (between two vectors) and a matrix product (between two matrices), and presented functions to compute these. Two methods for solving a system of N equations with N unknowns using matrix operations were presented. One method used the inverse of a matrix, and the other used matrix left division.

MATLAB Summary

This MATLAB summary lists and briefly describes all of the special characters, commands, and functions that were defined in this chapter:

Special Characters	
'	indicates a matrix transpose
*	matrix multiplication
\	matrix left division

Commands and Functions	
`det`	computes the determinate of a matrix
`dot`	computes the dot product of two vectors
`diag`	extracts the diagonal from a matrix, or generates a matrix with the input on the diagonal
`eye`	generates an identity matrix
`fliplr`	flips a matrix from left to right
`flipud`	flips a matrix from up to down
`inv`	computes the inverse of a matrix
`magic`	generates a magic square
`ones`	generates a matrix composed of ones
`zeros`	generates a matrix composed of zeros

KEY TERMS

conformable
determinant
diagonal matrix
dot product
identity matrix
ill-conditioned matrix
inverse
magic square
main diagonal
matrix left division
matrix multiplication
singular matrix
system of equations
transpose

PROBLEMS

1. Compute the dot product of the following pairs of vectors, and then show that

$$\mathbf{A} \cdot \mathbf{B} = \mathbf{B} \cdot \mathbf{A}$$

(a) $\mathbf{A} = [1 \quad 3 \quad 5]$, $\mathbf{B} = [-3 \quad -2 \quad 4]$
(b) $\mathbf{A} = [0 \; -1 \; -4 \; -8]$, $\mathbf{B} = [4 \; -2 \; -3 \; 24]$

2. Compute the total mass of the following components, using a dot product:

Component	Density, g/cm³	Volume, cm³
Propellant	1.2	700
Steel	7.8	200
Aluminum	2.7	300

3. Bomb calorimeters are used to determine the energy released during chemical reactions. The total heat capacity of a bomb calorimeter is defined as the sum of the product of the mass of each component and the specific heat capacity of each component. That is,

$$\text{CP} = \sum_{i=1}^{n} m_i C_i$$

where

m_i is the mass of each component, g;
C_i is the heat capacity of each component, J/gK; and
CP is the total heat capacity, J/K.

Find the total heat capacity of a bomb calorimeter with the following components:

Component	Mass, g	Heat Capacity, J/gK
Steel	250	0.45
Water	100	4.2
Aluminum	10	0.90

4. Compute the matrix product **A*B** of the following pairs of matrices:

(a) $\mathbf{A} = [12 \;\; 4; 3 \; -5]$, $\mathbf{B} = [2 \; 12; 0 \; 0]$
(b) $\mathbf{A} = [1 \; 3 \; 5; 2 \; 4 \; 6]$, $\mathbf{B} = [-2 \; 4; 3 \; 8; 12 \; -2]$

5. A series of experiments were performed with the bomb calorimeter from Problem 3. In each experiment, a different amount of water was used, as shown in the following table:

Experiment #	Mass of Water, g
1	110.0
2	100.0
3	101.0
4	98.6
5	99.4

Calculate the total heat capacity for the calorimeter for each of the experiments.

6. Given the array **A** = [−1 3; 4 2], raise each element of **A** to the second power. Raise **A** to the second power by matrix exponentiation. Explain why the answers are different.
7. Given the array **A** = [−1 3; 4 2], compute the determinant of `A`.
8. If **A** is conformable to **B** for addition, then a theorem states that $(\mathbf{A} + \mathbf{B})^{\mathrm{T}} = A^T + B^T$. Use MATLAB to test this theorem on the following matrices:

$$A = \begin{bmatrix} 2 & 12 & -5 \\ -3 & 0 & -2 \\ 4 & 2 & -1 \end{bmatrix} \qquad B = \begin{bmatrix} 4 & 0 & 12 \\ 2 & 2 & 0 \\ -6 & 3 & 0 \end{bmatrix}$$

9. Given that matrices **A**, **B**, and **C** are conformable for multiplication, the associative property holds; that is, **A(BC)** = **(AB)C**. Test the associative property using matrices **A** and **B** from Problem 8, along with matrix **C**:

$$\mathrm{C} = \begin{bmatrix} 4 \\ -3 \\ 0 \end{bmatrix}$$

10. Recall that not all matrices have an inverse. A matrix is singular (i.e., it does not have an inverse) if $|\mathrm{A}| = 0$. Test the following matrices using the determinant function to see if each has an inverse:

$$\mathbf{A} = \begin{bmatrix} 2 & -1 \\ 4 & 5 \end{bmatrix}, \quad \mathbf{B} = \begin{bmatrix} 4 & 2 \\ 2 & 1 \end{bmatrix}, \quad \mathbf{C} = \begin{bmatrix} 2 & 0 & 0 \\ 1 & 2 & 2 \\ 5 & -4 & 0 \end{bmatrix}$$

If an inverse exists, compute it.

11. Solve the following systems of equations using both the matrix left division and the inverse matrix methods:

(a)
$$\begin{aligned} -2x_1 + x_2 &= -3 \\ x_1 + x_2 &= 3 \end{aligned}$$

(b)
$$\begin{aligned} 10x_1 - 7x_2 + 0x_3 &= 7 \\ -3x_1 + 2x_2 + 6x_3 &= 4 \\ 5x_1 + x_2 + 5x_3 &= 6 \end{aligned}$$

(c)
$$\begin{aligned} x_1 + 4x_2 - x_3 + x_4 &= 2 \\ 2x_1 + 7x_2 + x_3 - 2x_4 &= 16 \\ x_1 + 4x_2 - x_3 + 2x_4 &= -15 \\ 3x_1 - 10x_2 - 2x_3 + 5x_4 &= -15 \end{aligned}$$

12. Time each method you used in Problem 11 for part c by using the **`clock`** function and the **`etime`** function, the latter of which measures elapsed time. Which method is faster, left division or inverse matrix multiplication?

```
t0 = clock;
(code to be timed)
etime(clock,t0)
```

13. In Example 6.4, we showed that the circuit shown in Figure 6.3 could be described by the following set of linear equations:

$$\begin{aligned} (R_2 + R_4)\,i_1 + (-R_2)\,i_2 + (-R_4)\,i_3 &= V_1 \\ (-R_2)\,i_1 + (R_1 + R_2 + R_3)\,i_2 + (-R_3)\,i_3 &= 0 \\ (-R_4)\,i_1 + (-R_3)\,i_2 + (R_3 + R_4 + R_5)\,i_3 &= 0 \end{aligned}$$

We solved this set of equations using the matrix inverse approach. Redo the problem, but this time use the left division approach.

14. **Amino Acids.** The amino acids in proteins contain molecules of oxygen (O), carbon (C), nitrogen (N), sulfur (S), and hydrogen (H), as shown in Table 6.1. The molecular weights for oxygen, carbon, nitrogen, sulfur, and hydrogen are as follows:

Oxygen	15.9994
Carbon	12.011
Nitrogen	14.00674
Sulfur	32.066
Hydrogen	1.00794

(a) Write a program in which the user enters the number of oxygen atoms, carbon atoms, nitrogen atoms, sulfur atoms, and hydrogen atoms in an amino acid. Compute and print the corresponding molecular weight. Use a dot product to compute the molecular weight.

(b) Write a program that computes the molecular weight of each amino acid in Table 6.1, assuming that the numeric information in this table is contained in a data file named **elements.dat**. Generate a new data file named **weights.dat** that contains the molecular weights of the amino acids. Use matrix multiplication to compute the molecular weights.

Table 6.1 Amino Acid Molecules

Amino Acid	O	C	N	S	H
Alanine	2	3	1	0	7
Arginine	2	6	4	0	15
Asparagine	3	4	2	0	8
Aspartic	4	4	1	0	6
Cysteine	2	3	1	1	7
Glutamic	4	5	1	0	8
Glutamine	3	5	2	0	10
Glycine	2	2	1	0	5
Histidine	2	6	3	0	10
Isoleucine	2	6	1	0	13
Leucine	2	6	1	0	13
Lysine	2	6	2	0	15
Methionine	2	5	1	1	11
Phenylanlanine	2	9	1	0	11
Proline	2	5	1	0	10
Serine	3	3	1	0	7
Threonine	3	4	1	0	9
Tryptophan	2	11	2	0	11
Tyrosine	3	9	1	9	11
Valine	2	5	1	0	11

CHAPTER

7 Symbolic Mathematics

Objectives

After reading this chapter, you should be able to

- create and manipulate symbolic variables,
- factor and simplify mathematical expressions,
- solve symbolic expressions,
- solve systems of equations, and
- determine the symbolic derivative of an expression and integrate an expression.

ENGINEERING ACHIEVEMENT: GEOLOCATION

In an earlier chapter, we discussed the use of the Global Positioning System (GPS) that is used by airliners, ships, trucks, and cars to accurately specify their location. This system is based on a constellation of 24 satellites that broadcasts position, velocity, and time information worldwide. Since GPS systems are not available for all applications, we also need to be able to rely on the equations for distance, velocity, and acceleration to help identify locations. For example, the characteristics of unpowered projectiles (such as bullets or flares that are used to signal a location) can be analyzed using these equations. In this chapter, we use symbolic mathematics to analyze the distance traveled and the time to impact for unpowered projectiles.

7.1 SYMBOLIC ALGEBRA

In addition to using numbers, we can use symbols to perform computations in MATLAB. This capability to manipulate mathematical expressions without using numbers can be very useful in solving certain types of engineering problems. A complete set of **symbolic functions** is available in the Symbolic Math Toolbox. The professional version of MATLAB includes the entire toolbox; a subset is included with the Student Edition.

Symbolic algebra is used to factor and simplify mathematical expressions, to determine solutions to equations, and to perform integration and differentiation of mathematical expressions. Additional capabilities (not discussed here) include linear algebra functions for determining inverses, determinants, eigenvalues, and canonical forms of symbolic matrices; variable precision arithmetic for numerically evaluating mathematical expressions to any specified accuracy; symbolic and numerical solutions to differential equations; and special mathematical functions that evaluate functions such as Fourier transforms. For more details on these additional symbolic capabilities, refer to the Help browser.

7.1.1 Symbolic Expressions

A **symbolic expression** is stored in MATLAB as a character string. Single quotes are used to define the symbolic expression, which is entered as the argument of the `sym` function. Consider these statements and corresponding display:

```
S = sym('x^2 - 2*y^2 + 3*a')
S =
x^2 - 2*y^2 + 3*a
```

Notice that symbolic results are not indented in the Command Window. All of the calculations we have performed in previous chapters return an indented result, indicating that they are double-precision, floating-point numbers. Notice in the Workspace Window that `S` is identified as a symbolic variable. Although **`x`**, **`y`**, and `a` are part of the symbolic expression `S`, they are not listed individually in the Workspace window.

To identify the variables in a symbolic expression or matrix, use the **`findsym`** function. This function returns the variables of its single argument in alphabetical order:

```
findsym(S)
ans =
a,x,y
```

Notice again that the results are not indented and that **`ans`** is listed in the Workspace window as a character variable.

You can create complex expressions in two different ways: by defining the entire expression all at once as we have just done or by defining each symbolic variable individually. You can use the **`sym`** function to define each variable one at a time, as shown here:

```
a = sym('a')
a =
a
```

You can also use the **`syms`** command to define multiple symbolic variables all at once:

```
syms a x y
```

Obviously, the second approach is quicker because it requires fewer statements. Your Workspace window should now reflect the fact that **a**, **x**, and **y** are symbolic variables.

HINT

Remember that you can adjust the workspace view with workspace options. Use **Choose Columns**, and check the columns you would like to be shown. Be sure to include Class in order to distinguish symbolic variables in this view.

Now you can use the symbolic variables to create new expressions:

```
S_new = x^2 - 2*y^2 + 3*a;
```

To substitute variables or expressions within a symbolic expression, use the **subs** command. When used with three arguments, the syntax of **subs** is

```
subs(S,old,new);
```

where **S** is a symbolic expression, **old** is a symbolic variable or string that represents a variable name, and **new** is a symbolic or numeric variable or expression. Multiple substitutions may be made by listing arguments within curly braces. For example, the following statement substitutes the variable **b** for variable **a** in the symbolic expression **S**:

```
subs(S,'a','b')
ans =
x^2 - 2*y^2 + 3*b
```

To substitute the variables **q** and **r** for the variables **x** and **y**, respectively, use this statement:

```
subs(S,{'x','y'},{'q','r'})
ans =
q^2 - 2*r^2 + 3*a
```

Note that neither of these commands stored the value of the new expression into **S**. For this reason, the first substitution was not reflected in the final answer after the second substitution.

HINT

If you define each variable separately using the **sym** command, you do not need the single quotes around the variable name when you use **subs**. For example, in **subs(S,x,'q')**, **x** has been defined as a symbolic variable, but **q** has not been defined.

We now want to create symbolic expressions **S1**, **S2**, **S3**, **S4**, and **S5** for the following mathematical expressions:

$$\mathbf{S1} = x^2 - 9$$

$$\mathbf{S2} = (x - 3)^2$$

$$\mathbf{S3} = \frac{x^2 - 3x - 10}{x + 2}$$

$$\mathbf{S4} = x^3 + 3x^2 - 13x - 15$$

$$\mathbf{S5} = 2x - 3y + 4x + 13b - 8y$$

Since **x** has already been defined as a symbolic variable, we can enter expressions **S1**, **S2**, **S3**, and **S4** directly as shown below:

```
S1 = x^2 - 9;
S2 = (x-3)^2;
S3 = (x^2-3*x-10)/(x+2);
S4 = x^3 + 3*x^2 - 13*x - 15;
```

We could also use statements with the following format:

```
S1 = sym('x^2 - 9');
```

We need to use the variables **x**, **y**, and **b** in expression **S5**. Since **b** has not been defined as a symbolic variable, we will either need to use the **sym** command or define **b** separately:

```
S5 = sym('2*x - 3*y + 4*x + 13*b - 8*y');
```

Also note that sometimes expressions do not print as we would have expected. For example, if we enter **S3** in the command window, the following is displayed:

```
S3
ans =
-(- x^2 + 3*x + 10)/(x + 2)
```

This expression is equivalent to the one entered for **S3**.

7.1.2 Symbolic Plotting

MATLAB includes a function called **ezplot** that generates a **symbolic plot**—a plot of a symbolic expression of one variable. The independent variable ranges by default over the interval $[-2\pi, 2\pi]$. A second form of **ezplot** allows the user to specify the range. If the variable contains a singularity (i.e., a point at which the expression is not defined), that point is not plotted. The syntax for the **ezplot** function is described as follows:

`ezplot(S)`	Generates a plot of **S** in the range $[-2\pi, 2\pi]$. **S** is assumed to be a function of one variable.
`ezplot(S,[xmin,xmax])`	Generates a plot of **S** in the range **[xmin,xmax]**. **S** is assumed to be a function of one variable.

We can use the symbolic expression **S4**, defined before, to demonstrate **ezplot**:

```
ezplot(S4)
```

This plot is shown in Figure 7.1. To change the title and labels, you could add these statements:

```
ezplot(S4),
   title('Plot of a second-order polynomial')
   xlabel('x'), ylabel('y')
```

7.1.3 Simplification of Mathematical Expressions

A number of functions are available for simplifying mathematical expressions by **collecting coefficients**, **expanding terms**, **factoring expressions**, or just making the expression simpler. A summary of these functions is as follows:

`collect(S)`	Collects coefficients of **S**.
`collect(S,'v')`	Collects coefficients of **S** with respect to the independent variable **'v'**.
`expand(S)`	Performs an expansion of **S**.
`factor(S)`	Returns the factorization of **S**.
`simple(S)`	Simplifies the form of **S** to a shorter form, if possible.
`simplify(S)`	Simplifies **S**.

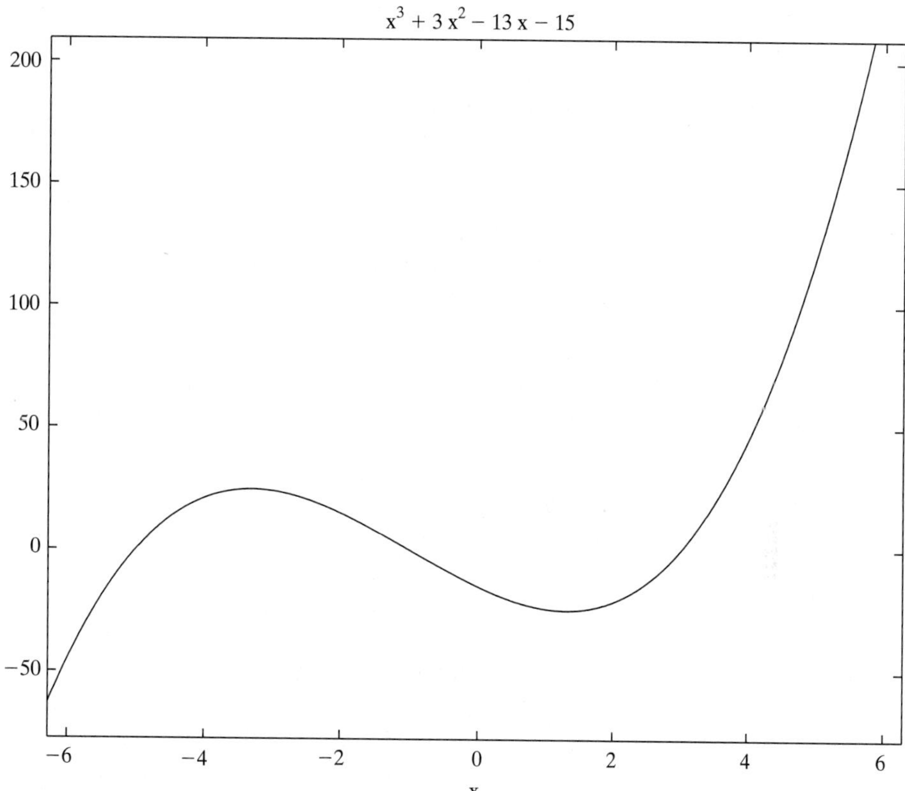

Figure 7.1
Plot of expression **S4**.

To illustrate these functions, we use the following symbolic expressions **S1**, **S2**, **S3**, **S4**, and **S5**:

```
factor(S1)
ans =
(x - 3)*(x + 3)


expand(S2)
ans =
x^2 - 6*x + 9


simplify(S3)
ans =
x - 5


factor(S4)
ans =
(x - 3)*(x + 5)*(x + 1)


collect(S5)
ans =
6*x + 13*b - 11*y
```

The **simple** function attempts to simplify a symbolic expression by using several different algebraic methods. Each method is displayed, along with its results. Finally, the shortest result is chosen as the answer.

As an example, we will use the **simple** function to find the simplest form of the following expression, which we had defined as **S3**:

$$\frac{x^2 - 3x - 10}{x + 2}$$

First, we create a symbolic expression **S3**, and then apply the **simple** function:

```
S3 = sym('(x^2-3*x-10)/(x+2)');
simple(S3)
```

The methods used and the various results will be displayed, with the recommended choice shown as the value for **ans**. For example, in this case, one of the outputs is the following:

```
radsimp: -(- x^2 + 3*x + 10)/(x + 2)
```

and the final output is:

```
ans =
x - 5
```

7.1.4 Operations on Symbolic Expressions

The standard arithmetic operations can be applied to symbolic expressions. Addition, subtraction, multiplication, division, and raising an expression to a power are performed by using the standard arithmetic operators. This feature can best be illustrated with examples. Before proceeding with the examples, we will create symbolic objects **S6**, **S7**, and **S8** for the following expressions:

$$\mathbf{S6} = \frac{1}{y - 3}$$

$$\mathbf{S7} = \frac{3y}{y + 2}$$

$$\mathbf{S8} = (y + 4)(y - 3)y$$

These symbolic objects can be defined with the following MATLAB statements:

```
S6 = 1/(y-3);
S7 = 3*y/(y+2);
S8 = (y+4)*(y-3)*y;
```

We now multiply the symbolic objects for **S6** and **S8**, and then use the **pretty** function to display the results in typeset form:

```
pretty(S6*S8)
y (y + 4)
```

Similarly, we can raise the symbolic expression **S7** to the third power and use **pretty** to print the results:

```
pretty(S7^3)
          3
    27 y
    --------
           3
    (y + 2)
```

The **poly2sym** function converts a numerical vector to a symbolic expression. The symbolic representation is the polynomial whose coefficients are listed in the vector. The default independent variable is *x*, but another variable may be named as a second argument to the function. The **sym2poly** function creates a coefficient vector from the symbolic representation of a polynomial.

First, we create a vector containing the coefficients of the polynomial:

```
V = [1,-4,0,2,45];
```

Then, we create a symbolic expression for the polynomial represented by **V**:

```
poly2sym(V)
```

The result is

```
ans =
x^4 - 4*x^3 + 2*x + 45
```

7.2 EQUATION SOLVING

Symbolic math functions can be used to solve a single equation, a system of equations, and differential equations. Brief descriptions of the functions for solving a single equation or a system of equations are as follows:

`solve(f)`	Solves the symbolic equation `f` for its symbolic variable. Solves the equation **f** = **0** for its symbolic variable if `f` is a symbolic expression. If there are multiple variables, MATLAB solves for `x`.
`solve(f1,...fn)`	Solves the system of equations represented by `f1,...fn`.
`solve(f,'y')`	Solves the symbolic equation `f`, for the variable `y`. You do not need single quotes around the second argument if it has been explicitly defined as symbolic.

To illustrate the use of the **solve** function, assume that the following equations have been defined:

```
eq1 = sym('x-3=4');
eq2 = sym('x^2-x-6');
eq3 = sym('x^2 + 2*x + 4 = 0');
eq4 = sym('3*x + 2*y - z = 10');
eq5 = sym('-x + 3*y + 2*z = 5');
eq6 = sym('x - y - z = -1');
```

Notice that in **eq1** the symbolic expression is an equation, but in **eq2** the expression is not set equal to anything. In this case, MATLAB will assume that the expression is equal to zero.

We now show a number of statements that use the **solve** function, along with their output:

```
solve(eq1)
ans =
7


solve(eq2)
ans =
3
-2
```

```
solve(eq3)
ans =
3^(1/2)*i - 1
-3^(1/2)*i - 1


solve(eq4)
ans =
z/3 - (2*y)/3 + 10/3


solve(eq4,y)
ans =
z/2 - (3*x)/2 + 5
```

When we attempt to solve the set of simultaneous equations, **eq4**, **eq5**, and **eq6**, the result is puzzling. It tells us that the results arc 1 × 1 symbolic variables, but it does not reveal the value of those variables. To force the results to be displayed, we must assign them variables names, as shown in this statement:

```
[X1, X2, X3,...Xn] = solve(f1,...fn);
```

The results are assigned alphabetically. For example, if the variables used in your symbolic expressions are **q**, **x**, and **p**, the results will be returned in the order **p**, **q**, **x**, independently of the names you have assigned for the results. Therefore, to solve the system of three equations we use the following:

```
[x, y, z] = solve(eq4,eq5,eq6)
x =
-2
y =
5
z =
-6
```

The result of the **solve** function is a symbolic variable, either **ans** or a user-defined name. If you want to use that result in a MATLAB expression that requires a double-precision, floating-point input, you can change the variable type with the **double** function. For example,

```
double(x)
```

changes **x** from a symbolic variable to a matrix variable.

The function for solving ordinary differential equations is **dsolve**, but it is not discussed in this text. For more information, consult the MATLAB help files.

EXAMPLE 7.1

FIND THE RANGE OF A FLARE

We can use the symbolic math capabilities of MATLAB to explore the equations representing the path followed by an unpowered projectile, such as a flare that is used to identify a location when GPS may not be available. We know from elementary physics that the distance a projectile travels horizontally is

$$d_x = v_0 t \cos(\theta)$$

and the distance traveled vertically is

$$d_y = v_0 t \sin(\theta) - \frac{1}{2}gt^2$$

where

v_0 is the velocity at launch,
t is time,
θ is the launch angle, and
g is the acceleration due to gravity.

Use these equations and MATLAB's symbolic capability to derive an equation for the distance the flare has traveled horizontally when it hits the ground (the range).

SOLUTION

1. Problem Statement

Find the range equation of an unpowered projectile.

2. Input/Output Description

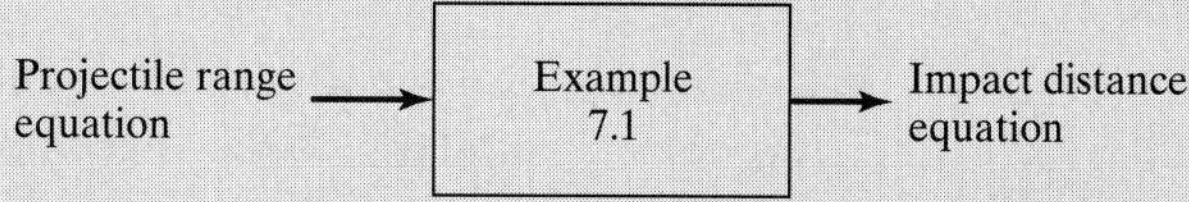

3. Hand Example

$$d_y = v_0 t \sin(\theta) - \frac{1}{2}gt^2 = 0$$

Rearrange to give

$$v_0 t \sin(\theta) = \frac{1}{2}gt^2$$

Divide by t and solve:

$$t = \frac{v_0 \sin(\theta) \times 2}{g}$$

Now substitute this expression for t into the horizontal distance formula:

$$d_x = v_0 t \cos(\theta)$$

$$\text{range} = v_0\left(\frac{v_0 \sin\,(\theta) \times 2}{g}\right)\cos\,(\theta)$$

We know from trigonometry that $\sin\theta\,\cos\theta$ is the same as $\sin\,(2\theta)$, which would allow a further simplification if we wanted it.

(*continued*)

4. MATLAB Solution

We have a symbolic equation for the vertical distance that the flare travels, and for the corresponding horizontal distance. We can solve the vertical distance equation for the impact time, which is where the vertical distance is zero. We can then substitute that time in the horizontal equation to give the range of the flare. Since the distance equation is a quadratic equation, there are two solutions: The first solution is zero, so the second solution is the one in which we are interested.

```
%----------------------------------------------------------------
% Example 7_1 - This program determines the equation
% for the range of a projectile in terms of the
% initial velocity and the angle of flight.
%
clear, clc
%
syms v0 t theta g
%
% Determine the impact time from distance.
Distance_y = v0*t*sin(theta) - 1/2*g*t^2;
Distance_x = v0*t*cos(theta);
impact_time = solve(Distance_y,t)
%
% Display the equation for the range.
impact_distance = subs(Distance_x,t,impact_time(2))
%----------------------------------------------------------------
```

5. Testing

Compare the MATLAB solution with the hand solution. Both approaches give the same result and agree with the hand example.

```
impact_distance =
(2*v0^2*cos(theta)*sin(theta))/g
```

MATLAB can simplify the result, although this is already fairly simple. We use the **simple** command to demonstrate all the possibilities:

```
simple(impact_distance)
```

The final answer from this command is the following:

```
ans =
v0^2*sin(2*theta)/g
```

EXAMPLE 7.2

PLOT THE FLARE RANGE AS A FUNCTION OF ANGLE

In Example 7.1, we used MATLAB's symbolic capabilities to find an equation for the range of the flare as a function of the initial velocity at launch and the launch angle. We now use MATLAB's symbolic capabilities to create a plot showing the range traveled for angles from 0 to $\pi/2$. Assume an initial velocity of 100 m/s and an acceleration due to gravity of 9.8 m/s^2.

SOLUTION

1. Problem Statement

Plot the range as a function of launch angle.

2. Input/Output Description

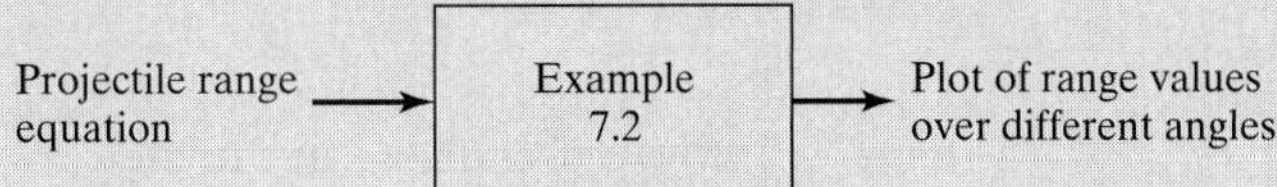

3. Hand Example

From Example 7.1, we have the equation for the range:

$$range \;=\; v_0^2 \frac{\sin\,(2\theta)}{g}$$

Using this equation, we can easily calculate a few data points:

angle, radians	range, m
0	0
$\pi/6$	884
$\pi/4$	1020
$\pi/3$	884
$\pi/2$	0

The range appears to increase with increasing angle, and then decrease back to zero when the flare gun is pointed straight up.

4. MATLAB Solution

From Example 7.1, we have a symbolic equation that represents the impact distance. We can substitute the numerical values of launch velocity and acceleration due to gravity into the equation, and then plot the resulting equation.

```
%-----------------------------------------------------------
% Example 7_2 - This program uses the equation for
% the range of a projectile to generate a plot of
% range over the different angles of flight.
%
clear, clc
%
syms v0 t theta g
%
% Determine the equation for the range for
% initial velocity of 100 m/s and acceleration
% due to gravity.
```

(continued)

```
impact_distance = v0^2*sin(2*theta)/g;
impact_100 = subs(impact_distance,{v0,g},{100,9.8});
ezplot(impact_100,[0, pi/2]),
   title('Maximum Projectile Distance Traveled'),
   xlabel('angle, radians'),
   ylabel('height, m')
%-----------------------------------------------------------
```

5. Testing

The plot generated is shown in Figure 7.2. The values that we computed in the hand example match those in the plot.

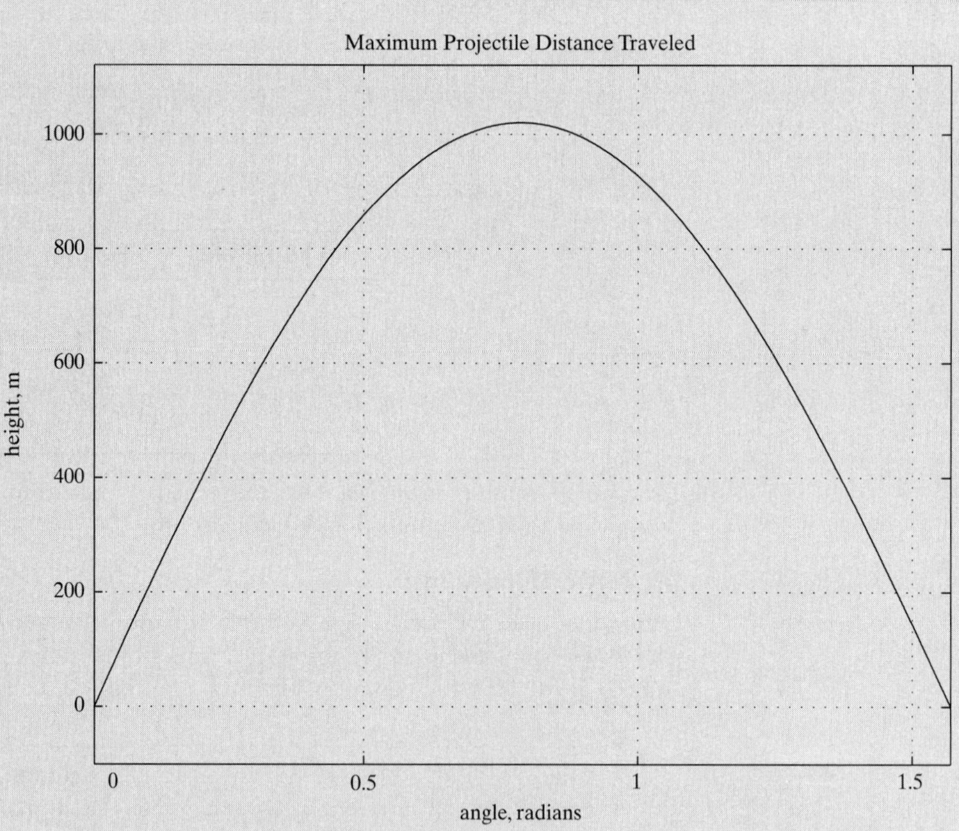

Figure 7.2
The range of a projectile depends on the initial velocity and the launch angle.

7.3 DIFFERENTIATION AND INTEGRATION

The operations of **differentiation** and **integration** are used extensively in solving engineering problems. In this section, we discuss the differentiation and integration of symbolic expressions.

7.3.1 Differentiation

The **diff** function is used to determine the symbolic derivative of a symbolic expression. There are four forms in which this function can be used to perform symbolic differentiation:

`diff(f)`	Returns the derivative of the expression `f` with respect to the default independent variable.
`diff(f,'t')`	Returns the derivative of the expression `f` with respect to the variable `t`.
`diff(f,n)`	Returns the `n`th derivative of the expression `f` with respect to the default independent variable.
`diff(f,'t',n)`	Returns the `n`th derivative of the expression `f` with respect to the variable `t`.

We now present several examples using the **diff** function for symbolic differentiation. First, we define the following symbolic expressions:

```
S1 = sym('6*x^3-4*x^2+b*x-5');
S2 = sym('sin(a)');
S3 = sym('(1-t^3)/(1+t^4)');
```

The following table shows function references and their corresponding values:

Reference	Function Value
`diff(S1)`	`18*x^2 - 8*x + b`
`diff(S1,2)`	`36*x - 8`
`diff(S1,'b')`	`x`
`diff(S2)`	`cos(a)`
`diff(S3)`	`(4*t^3*(t^3-1))/(t^4+1)^2 - (3*t^2)/(1+t^4)`
`simplify(diff(S3))`	`-(t^2*(-t^4+4*t+3))/(t^4+1)^2`

7.3.2 Integration

The **int** function is used to integrate a symbolic expression **f**. This function attempts to find the symbolic expression **F** such that **diff(F)** = **f**. It is possible that the integral (or antiderivative) may not exist in closed form or that MATLAB cannot find the integral. In such cases, the function will return the unevaluated command. The **int** function can be used in the following forms:

`int(f)`	Returns the integral of the expression `f` with respect to the default independent variable.
`int(f,'t')`	Returns the integral of the expression `f` with respect to the variable `t`.
`int(f,a,b)`	Returns the integral of the expression `f` with respect to the default independent variable evaluated over the interval `[a,b]`, where `a` and `b` are numeric expressions.
`int(f,'t',a,b)`	Returns the integral of the expression `f` with respect to the variable `t` evaluated over the interval `[a,b]`, where `a` and `b` are numeric expressions.
`int(f,'m','n')`	Returns the integral of the expression `f` with respect to the default independent variable evaluated over the interval `[m,n]`, where `m` and `n` are symbolic expressions.

We now present several examples that use the **int** function for symbolic integration. First, we define the following symbolic expressions:

```
S1 = sym('6*x^3-4*x^2+b*x-5');
S2 = sym('sin(a)');
S3 = sym('sqrt(x)');
```

The following table shows function references and their corresponding values:

Reference	Function Value
`int(S1)`	`(3*x^4)/2 - (4*x^3)/3 + (b*x^2)/2 - 5*x`
`int(S2)`	`-cos(a)`
`int(S3)`	`(2*x^(3/2))/3`
`int(S3,'a','b')`	`(2*b^(3/2))/3 - (2*a^(3/2))/3`
`int(S3, 0.5, 0.6)`	`(2*15^(1/2))/25 - 2^(1/2)/6`
`double(int(S3, 0.5, 0.6))`	`0.0741`

EXAMPLE 7.3

DETERMINE ANGLE FOR MAXIMUM RANGE

Using the equations presented in Examples 7.1 and 7.2, use MATLAB's symbolic capability to find the angle where the maximum range occurs, and to find the maximum range, assuming the velocity at launch is 100 m/s and that the acceleration is due to the Earth's gravity.

SOLUTION

1. Problem Statement

Find the positive angle where the maximum range occurs and find the corresponding range.

2. Input/Output Description

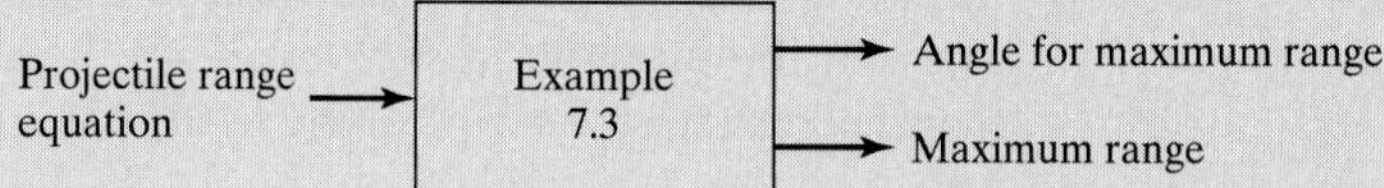

3. Hand Example

From the graph shown in Figure 7.2, the maximum range appears to occur at a launch angle of approximately 0.7 or 0.8 radian, and the maximum height appears to be approximately 1000 m.

4. MATLAB Solution

Recall that the symbolic expression for the impact distance with v_0 and g defined as 100 m/s and 9.8 m/s^2, respectively, is

```
impact_100 =
10000*sin(2*theta)/9.8
```

From the graph, we can see that the maximum distance occurs when the slope is equal to zero. The slope is the derivative of **impact_100**, so we need to set the derivative equal to zero and solve. Since MATLAB automatically assumes that an expression is equal to zero, we can use the following equation to find the angle where the maximum height occurs:

```
max_angle = solve(diff(impact_100))
max_angle =
pi/4
```

This can be substituted into the expression for the range:

```
max_distance = subs(impact_100,theta,max_angle)
max_distance =
50000/49
```

(Note that 10000*sin(2*pi/4)/9.8 = 10000/9.8 = 50000/49.) To change the result to a decimal representation, we need to use the **double** function:

```
double(max_distance)
ans =
     1.0204e+03
```

5. Testing

The hand solution suggested an angle of 0.7 to 0.8 radians. The calculated value of pi/4 is equal to 0.785 radians. The estimated maximum height was approximately 1000 m, compared to the calculated 1024 m. Try the problem again with different initial launch velocities. As the launch velocity increases, so should the maximum height.

SUMMARY

In this chapter, we presented MATLAB's functions for performing symbolic mathematics. Examples were given to illustrate the simplification of expressions, the evaluation of operations with symbolic expressions, and the derivation of symbolic solutions to equations. In addition, we presented the MATLAB functions for determining the symbolic derivatives and integrals of expressions.

MATLAB Summary

This MATLAB summary lists and briefly describes all of the special characters, commands, and functions that were defined in this chapter.

Special Character	
'	used to enclose a symbolic expression

Commands and Functions	
`collect`	collects coefficients of a symbolic expression
`diff`	differentiates a symbolic expression
`double`	changes a symbolic variable into a double-precision, floating-point variable
`expand`	expands a symbolic expression
`ezplot`	generates a plot of a symbolic expression
`factor`	factors a symbolic expression
`findsym`	finds symbolic variables in a symbolic expression
`int`	integrates a symbolic expression
`poly2sym`	converts a vector to a symbolic polynomial
`pretty`	prints a symbolic expression in typeset form
`simple`	shortens a symbolic expression
`simplify`	simplifies a symbolic expression
`solve`	solves an equation
`subs`	replaces variables in a symbolic expression
`sym`	defines a symbolic expression
`sym2poly`	converts a symbolic expression to a coefficient vector

KEY TERMS

collecting coefficients
differentiation
expanding terms
factoring expression
integration
symbolic algebra
symbolic expression
symbolic function
symbolic plot

PROBLEMS

1. Create symbolic objects **S1** and **S2** for the following expression:

$$\mathbf{S1} = (x - 1)^2 + 2x - 1$$
$$\mathbf{S2} = x$$

2. Execute the **`simple`** function using **`S1`** as an argument. Which simplification method succeeds in finding the simplest expression for **`S1`**?
3. What is the result of the symbolic division **`S2/S1`**? What is the result of **`factor(S2/S1)`**?
4. Solve the equation

$$\frac{x - 1}{x^2 + 4} = 2.$$

5. Define a symbolic variable for each of the equations, and use MATLAB's symbolic capability to solve for each unknown.

$$\begin{aligned} x_1 - x_2 - x_3 - x_4 &= 5 \\ x_1 + 2x_2 + 3x_3 + x_4 &= -2 \\ 2x_1 + 2x_3 + 3x_4 &= 3 \\ 3x_1 + x_2 + 2x_4 &= 1 \end{aligned}$$

6. Compare the amount of time it takes to solve Problem 5 with left division and with symbolic math, using the **`tic`** and **`toc`** functions:

```
tic
(code to be timed)
toc
```

EQUATIONS

Determine the first and second derivatives of the following functions, using MATLAB's symbolic functions:

7. $g(x) = x^3 - 5x^2 + 2x + 8$
8. $g_2(x) = (x^2 + 4x + 4)*(x - 1)$
9. $g_3(x) = (x^2 - 2x + 2)/(10x - 24)$
10. $g_4(x) = (x^5 - 4x^4 - 9x^3 + 32)^2$

INTEGRALS

Use MATLAB's symbolic functions to determine the values of the following integrals:

11. $\int_{0.5}^{0.6} |x|dx$

12. $\int_{0}^{1} |x|dx$

13. $\int_{-1}^{-0.5} |x|dx$

14. $\int_{-0.5}^{0.5} |x|dx$

WEATHER BALLOONS

Assume that the following polynomial represents the altitude in meters during the first 48 hours after the launch of a weather balloon:

$$h(t) = -0.12t^4 + 12t^3 - 380t^2 + 4100t + 220$$

Assume that the units of t are hours.

15. Use MATLAB to determine the equation for the velocity of the weather balloon, using the fact that the velocity is the derivative of the altitude.

16. Use MATLAB to determine the equation for the acceleration of the weather balloon, using the fact that acceleration is the derivative of the velocity or the second derivative of the altitude.
17. Use MATLAB to determine when the balloon hits the ground. Because $h(t)$ is a fourth-order polynomial, there will be four answers. However, only one answer will be physically meaningful.
18. Use MATLAB's symbolic plotting capability to create plots of altitude, velocity, and acceleration from time zero until the balloon hits the ground. You'll need three separate plots, since altitude, velocity, and acceleration all have different units.
19. Determine the maximum height reached by the balloon. Use the fact that the velocity of the balloon is zero at the maximum height.

WATER FLOW

Assume that water is pumped into an initially empty tank. It is known that the rate of flow of water into the tank at time t (in seconds) is $50 - t$ liters per second. The amount of water Q that flows into the tank during the first x seconds can be shown to be equal to the integral of the expression $(50 - t)$ evaluated from 0 to x seconds.

20. Determine a symbolic equation that represents the amount of water in the tank after x seconds.
21. Determine the amount of water in the tank after 30 seconds.
22. Determine the amount of water that flowed into the tank between 10 seconds and 15 seconds after the flow was initiated.

ELASTIC SPRING

Consider a spring with the left end held fixed and the right end free to move along the x-axis. We assume that the right end of the spring is at the origin $x = 0$, when the spring is at rest. When the spring is stretched, the right end of the spring is at some new value of x that is greater than zero. When the spring is compressed, the right end of the spring is at some value that is less than zero. Assume that a spring has a natural length of 1 ft and that a force of $10\,\text{lb}_f$ is required to compress the spring to a length of 0.5 ft. It can then be shown that the work, in ft/lb_f done to stretch the spring from its natural length to a total of n ft, is equal to the integral of $20x$ over the interval from 0 to $n - 1$.

23. Use MATLAB to determine a symbolic expression that represents the amount of work necessary to stretch the spring to a total length of n ft.
24. What is the amount of work done to stretch the spring to a total of 2 ft?
25. If the amount of work exerted is $25\,\text{ft}/\text{lb}_f$, what is the length of the stretched spring?

CHAPTER

8 Numerical Techniques

Objectives

After reading this chapter, you should be able to

- perform linear and cubic-spline interpolations,
- calculate the best-fit straight line and polynomial to a set of data points,
- use the basic fitting tool, and
- perform numerical integration and differentiation.

ENGINEERING ACHIEVEMENT: DYNAMIC FLUID FLOW

The dynamics of fluid flow are difficult to model, but developing models to estimate these dynamics is important in solving many problems involving liquids or gases. For example, the design and construction of the Alaska pipeline presented numerous engineering challenges, and part of those challenges involved fluid flow. One problem that had to be addressed was how to protect the permafrost (the perennially frozen subsoil in Arctic or subarctic regions) from the heat of the pipeline itself. The oil flowing in the pipeline is warmed by pumping stations and by friction from the walls of the pipe; as a result, the supports holding the pipeline have to be insulated or even cooled to keep them from melting the permafrost at their bases. Many such physical processes are difficult to predict theoretically, but equations representing their behavior can be modeled using numerical techniques such as the ones presented in this chapter.

8.1 INTERPOLATION

Interpolation is a technique by which we estimate a variable's value between two known values. There are a number of different techniques for this, but in this section we present the two most common types of interpolation: linear interpolation and cubic-spline interpolation. In both techniques, we assume that we have a set of data points which represents a set of *xy*-coordinates for which *y* is a function of *x*; that is,

$y = f(x)$. We then have a value of x that is not part of the data set for which we want to find the y value.

8.1.1 Linear Interpolation

Linear Interpolation is one of the most common techniques for estimating data values between two given data points. With this technique, we assume that the function between the points can be estimated by a straight line drawn between the points. (See the top plot in Figure 8.1.) If we find the equation of a straight line defined by the two known points, we can find y for any value of x. The closer together the points are, the more accurate our approximation is likely to be. We could also use this equation to extrapolate points past our collected data, but generally we assume that valid estimates must be within the range of the data points.

8.1.2 Cubic-Spline Interpolation

A **cubic spline** is a smooth curve constructed to go through a set of points. The curve between each pair of points is a third-degree polynomial (which has the general form $a_0x^3 + a_1x^2 + a_2x + a_3$) which is computed so that it provides a smooth curve between the two points and a smooth transition from the third-degree polynomial between the previous pair of points. (See the bottom plot in Figure 8.1.)

8.1.3 interp1 Function

The MATLAB function that performs interpolation, **`interp1`**, has two forms. Each form assumes that vectors **`x`** and **`y`** contain the original data values and that another vector **`x_new`** contains the new point or points for which we want to compute

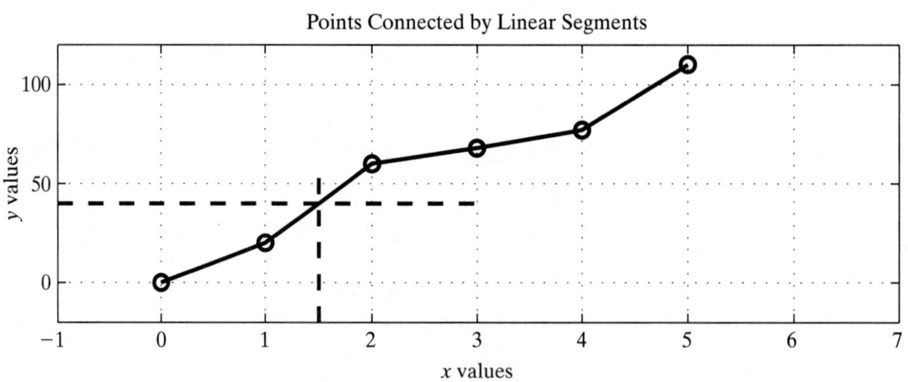

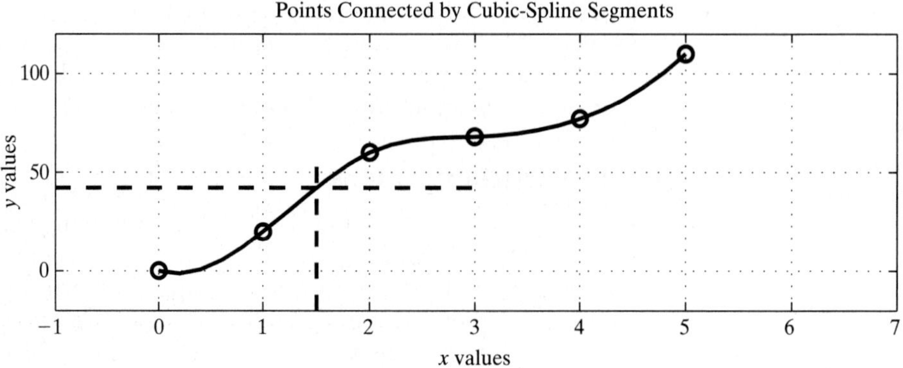

Figure 8.1
Top: Linear interpolation connects the points with a straight line segment, Bottom: Cubic-spline interpolation connects the points with cubic equations between each pair of points.

interpolated **y_new** values. (The **x** values should be in ascending order, and the **x_new** values should be within the range of the **x** values.) These forms are demonstrated in the following examples.

HINT

The last character in the function name **interp1** is the number one, "1". Depending on the font, it may look like the letter "l".

The points in Figure 8.1 were generated with the following commands:

```
x = 0:5;
y = [0,20,60,68,77,110];
```

Suppose we would like to find the value for y that corresponds to $x = 1.5$. Since 1.5 is not one of the elements in the **x** vector, we can perform an interpolation using the **interp1** function as shown in this statement:

```
interp1(x, y, 1.5)
ans =
     40
```

We can see from Figure 8.1 that this answer corresponds to a linear interpolation between the **x**, **y** points at (1,20) and (2,60). The function **interp1** defaults to linear interpolation unless otherwise specified.

If, instead of a single new **x** value, we define an array of new **x** values, the function returns an array of new **y** values:

```
new_x = 0:0.2:5;
new_y = interp1(x,y,new_x);
```

The new calculated points are plotted in Figure 8.2. They all fall on a straight line connecting the original data points. The commands to generate the graph are

```
plot(x,y,new_x,new_y,'o'),
    axis([-1,7,-20,120]),
    title('Linear Interpolation Plot'),
    xlabel('x values'),ylabel('y values'),grid
```

If we wish to use a cubic-spline interpolation approach, we must add a fourth argument to the **interp1** function. The argument must be a string. The choices are the following:

'nearest'	nearest neighbor interpolation
'linear'	linear interpolation—which is the default
'spline'	piecewise cubic-spline interpolation
'pchip'	shape-preserving piecewise cubic interpolation
'cubic'	same as **'pchip'**
'v5cubic'	the cubic interpolation from MATLAB 5 (a previous version of MATLAB), which does not extrapolate and uses **'spline'** if X is not equally spaced.

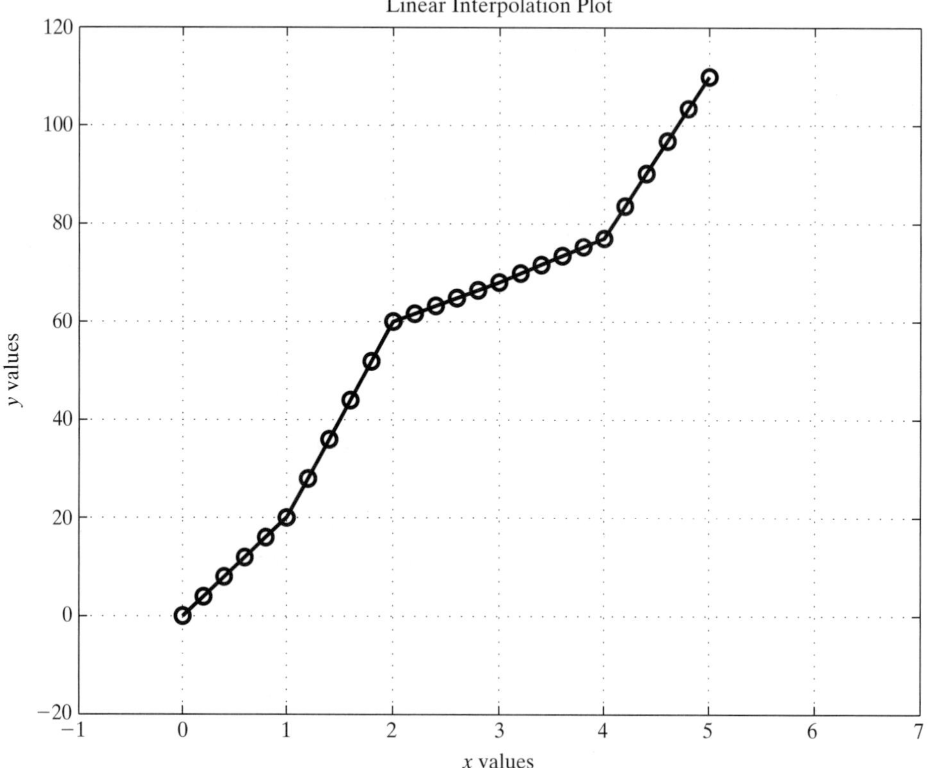

Figure 8.2
Example of linear interpolation.

To find the value of *y* where *x* is equal to 1.5 using a cubic spline, we can use the following:

```
interp1(x,y,1.5,'spline')
ans =
     42.2083
```

Referring to Figure 8.1, we see that this corresponds to our graph of a cubic spline. To generate a vector of new **y** values, we use the same procedure as before:

```
new_y = interp1(x,y,new_x,'spline');
```

The results are plotted in Figure 8.3. The original points are connected with a straight line. The curved plot is constructed from the calculated points. (Remember, all MATLAB plots are constructed of straight line segments, but these are close enough together to approximate a curve.)

The commands to generate Figure 8.3 are

```
plot(x,y,'x',new_x,new_y,'-o'),
   axis([-1,7,-20,120]),
   title('Cubic-Spline Interpolation Plot'),
   xlabel('x values'), ylabel('y values'),grid
```

MATLAB provides two-dimensional **(interp2)** and three-dimensional **(interp3)** interpolation functions, which are not discussed here. Refer to the help feature for more information.

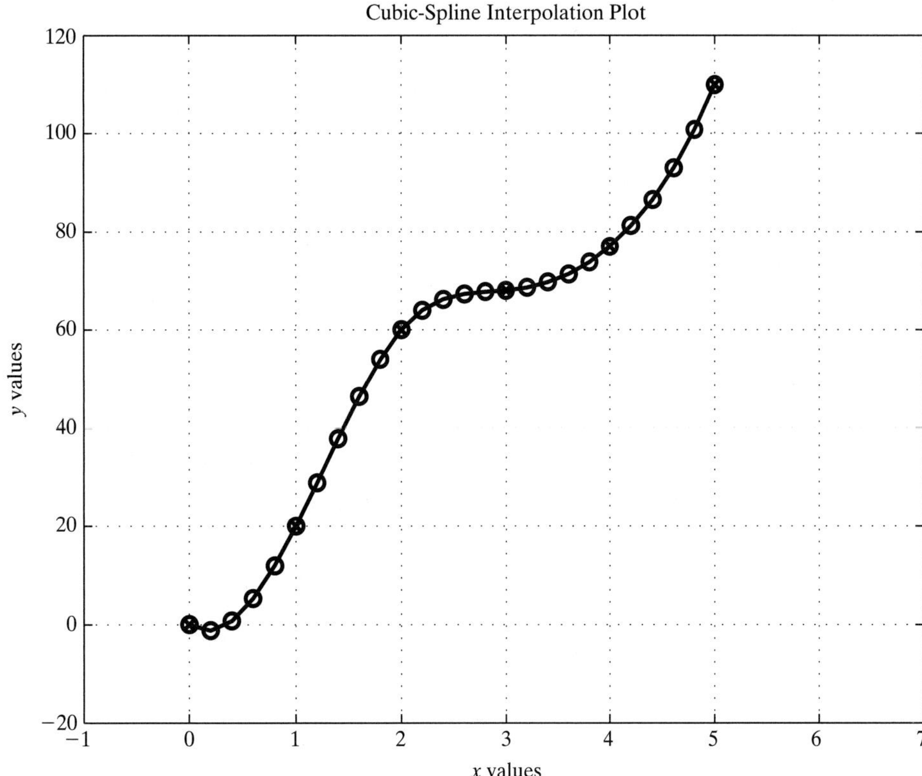

Figure 8.3
Example of cubic-spline interpolation.

EXAMPLE 8.1

INTERPOLATION USING STEAM TABLES

The subject of thermodynamics makes extensive use of tables. Although many properties can be described by fairly simple equations, others either are poorly understood, or the equations describing their behavior are very complicated. It is much easier just to tabulate the data. For example, consider the data for steam at 0.1 MPa (mega Pascals), or approximately 1 atmosphere which is the pressure at sea level, given in Table 8.1. These data could be used to analyze the geysers shown in Figure 8.4. Use linear interpolation to determine the internal energy at 215°C, and use linear interpolation to determine the temperature if the internal energy is 2600 kJ/kg.

SOLUTION

1. Problem Statement

Using linear interpolation, find the internal energy of steam at 215°C. Using linear interpolation, also find the temperature if the internal energy is 2600 kJ/kg.

(continued)

Table 8.1 Internal Energy as a Function of Temperature at 0.1 MPa

Temperature (°C)	Internal Energy kJ/kg
100	2506.7
150	2582.8
200	2658.1
250	2733.7
300	2810.4
400	2967.9
500	3131.6

Data from Joseph H. Keenan, Fredrick G. Keyes, Philip G. Hill, and Joan G. Moore, *Steam Tables: Thermodynamic Properties of Water Including Vapor, Liquid & Solid Phases* (New York: John Wiley and Sons, 1978).

Figure 8.4
Geysers spray high temperature and high-pressure water and steam.

2. Input/Output Description

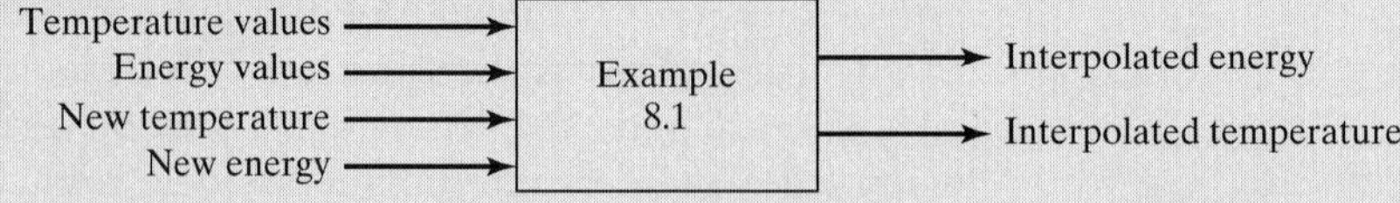

3. Hand Example

In the first part of the problem, we need to find the internal energy at 215°C. The table includes values at 200°C and 250°C. First, we need to find what fraction of the way between 200°C and 250°C the value for 215°C falls:

$$\frac{215 - 200}{250 - 200} = 0.30$$

If we model the relationship between temperature and internal energy as linear, the internal energy should also be 30 percent of the distance between the tabulated values:

$$0.30 = \frac{U - 2658.1}{2733.7 - 2658.1}$$

Solving for U, the interpolated internal energy value, gives

$$U = 2680.78 \text{ kJ/kg}$$

Using a similar process, if the internal temperature is 2600 kJ/kg, we can compute the temperature to be 161.4210°C.

4. MATLAB Solution

We create the solution in an M-file.

```
%-------------------------------------------------------------
%   Example 8_1 - This program uses linear interpolation
%   to find a temperature value given the internal energy,
%   and to find the internal energy given a temperature.
%
clear, clc
%
T = [100, 150, 200, 250, 300, 400, 500];
energy = [2506.7, 2582.8, 2658.1, 2733.7, 2810.4, ...
         2967.9, 3131.6];
new_energy = interp1(T, energy, 215)
newT = interp1(energy, T, 2600)
%-------------------------------------------------------------
```

5. Testing

The output of the program is shown below:

```
new_energy =
            2.6808e+03
newT =
      161.4210
```

The MATLAB result matches the hand result. This approach could be used for any of the properties tabulated in the Steam Tables. The JANAF tables are a similar source of thermodynamic properties published by the National Institute of Standards and Technology.

EXAMPLE 8.2

EXPANDING THE STEAM TABLES

Electric power plants use steam as a "working fluid" (see Figure 8.5). The science of thermodynamics makes extensive use of tables when systems such as a power plant are analyzed. Depending on the system of interest you may only need a portion of the table, such as Table 8.2.

(*continued*)

Figure 8.5
Power plants use steam as a "working fluid."

Table 8.2 Properties of Superheated Steam

Properties of Superheated Steam at 0.1 MPa (Approximately 1 ATM)			
Temperature (°C)	**Specific Volume (m3/kg)**	**Internal Energy (kJ/kg)**	**Enthalpy (kJ/kg)**
100	1.6958	2506.7	2676.2
150	1.9364	2582.8	2776.4
200	2.172	2658.1	2875.3
250	2.406	2733.7	2974.3
300	2.639	2810.4	3074.3
400	3.103	2967.9	3278.2
500	3.565	3131.6	3488.1

Data from Joseph H. Keenan, Fredrick G. Keyes, Philip G. Hill, and Joan G. Moore, *Steam Tables: Thermodynamic Properties of Water Including Vapor, Liquid & Solid Phases* (New York: John Wiley and Sons, 1978).

Notice that this table is spaced at 50-degree intervals at first, and then at 100-degree intervals. Assume that you have a project that requires you to use this table, and you would prefer not to have to perform a linear interpolation every time you use it. Use MATLAB to create a table, applying linear interpolation, with a temperature spacing of 25°C.

SOLUTION

1. Problem Statement

Find the specific volume, internal energy, and enthalpy every 25 degrees.

2. Input/Output Description

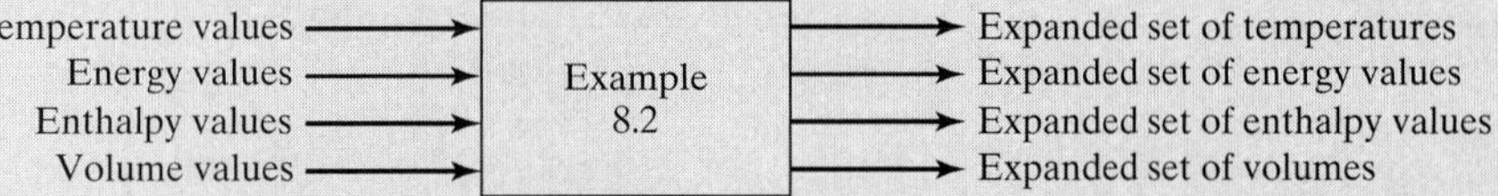

3. Hand Example

We will perform the calculations at 225°C, and find the corresponding energy, U:

$$\frac{225 - 200}{250 - 200} = 0.50 \text{ and } 0.50 = \frac{U - 2658.1}{2733.7 - 2658.1}$$

Solving for U gives

$$U = 2695.9 \text{ kJ/kg}$$

We can use this same calculation to confirm the calculations in the table we create.

4. MATLAB Solution

We create the MATLAB solution in an M-file.

```
%-------------------------------------------------------------
%  Example 8_2 - This program interpolates values
%  from a steam table to give an expanded table.
%
clear, clc
%
T = [100, 150, 200, 250, 300, 400, 500]';
energy = [2506.7, 2582.8, 2658.1, 2733.7, 2810.4,...
          2967.9, 3131.6]';
enthalpy = [2676.2, 2776.4, 2875.3, 2974.3, 3074.3,...
            3278.2, 3488.1]';
volume = [1.6958, 1.9364, 2.172, 2.406, 2.639,...
          3.103, 3.565]';
properties =[volume,energy,enthalpy];
newT = [100:25:500]';
newprop = interp1(T,properties, newT);
disp('Steam Properties at 0.1 MPa')
disp('Temp      Specific     Internal     Enthalpy')
disp('          Volume       Energy')
disp('C         m^3/kg       kJ/kg        kJ/kg')
fprintf('%3.0f %10.4f %11.1f %11.1f \n',...
       [newT, newprop]')
%-------------------------------------------------------------
```

5. Testing

The following table is printed by the MATLAB program:

(*continued*)

```
Steam Properties at 0.1 MPa
Temp       Specific          Internal          Enthalpy
           Volume            Energy
C          m^3/kg            kJ/kg             kJ/kg
100         1.6958            2506.7            2676.2
125         1.8161            2544.8            2726.3
150         1.9364            2582.8            2776.4
175         2.0542            2620.4            2825.9
200         2.1720            2658.1            2875.3
225         2.2890            2695.9            2924.8
250         2.4060            2733.7            2974.3
275         2.5225            2772.1            3024.3
300         2.6390            2810.4            3074.3
325         2.7550            2849.8            3125.3
350         2.8710            2889.2            3176.2
375         2.9870            2928.5            3227.2
400         3.1030            2967.9            3278.2
425         3.2185            3008.8            3330.7
450         3.3340            3049.8            3383.1
475         3.4495            3090.7            3435.6
500         3.5650            3131.6            3488.1
```

The MATLAB result matches the hand result. Now that we know the program works, we can create more extensive tables by changing the definition of **newT** from

```
newT = [100:25:500]';
```

to a vector with a smaller temperature increment, such as

```
newT =[100:1:500]';
```

8.2 CURVE FITTING: LINEAR AND POLYNOMIAL REGRESSION

Assume that we have a set of data points collected from an experiment. After plotting the data points, we find that they generally fall in a straight line. However, if we were to try to draw a straight line through the points, only a couple of the points would probably fall exactly on the line. A least-squares curve fitting method could be used to find the straight line that is the closest to the points, by minimizing the squared distance from each point to the straight line. Although this line can be considered a **best fit** to the data points, it is possible that none of the points would actually fall on the line of best fit. (Note that this method is very different from interpolation, because the curves used in linear interpolation and cubic-spline interpolation actually contained all of the original data points.) In this section, we first discuss fitting a straight line to a set of data points, and then we discuss fitting a polynomial to a set of data points.

8.2.1 Linear Regression

Linear regression is the name given to the process that determines the linear equation which is the best fit to a set of data points, in terms of minimizing the sum of the squared distances between the line and the data points. To understand this process, we first consider the following set of data values:

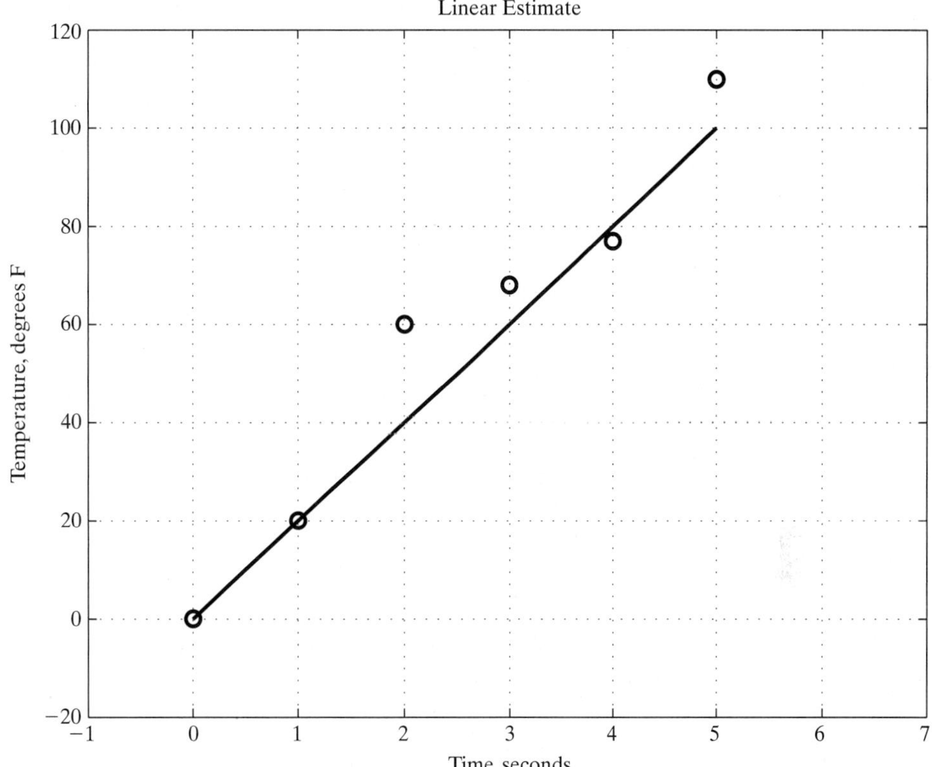

Figure 8.6
A linear estimate to a set of data points.

```
x = 0:5;
y = [0,20,60,68,77,110];
```

If we plot these points, it appears that a good estimate of a line through the points is $y = 20x$, as shown in Figure 8.6.

The following commands were used to generate the plot:

```
y2 = 20*x;
plot(x,y,'o',x,y2),
    axis([-1,7,-20,120]),
    title('Linear Estimate'),
    xlabel('Time,seconds'),
    ylabel('Temperature,degrees F'), grid
```

Looking at the plot, we can see that the first two points appear to fall exactly on the line, but the other points are off by varying amounts. To compare the quality of the fit of this line to other possible estimates, we find the difference between the actual y value and the value calculated from the estimate (in this case, $y = 20x$,). (These values are listed in Table 8.3.)

If we sum the differences, some of the positive and negative values would cancel each other out and give a sum that is smaller than it should be. To avoid this problem, we could add the absolute value of the differences, or we could square them. The least squared technique uses the squared values. Therefore, the measure of the quality of the fit of this linear estimate is the sum of the squared distances between

Table 8.3 Difference between Actual and Calculated Values

x	y (Actual)	y2 (Calculated)	Difference y - y2
0	0	0	0
1	20	20	0
2	60	40	20
3	68	60	8
4	77	80	-3
5	110	100	10

the points and the linear estimates. This sum, also called a **residual sum**, can be computed with the following command:

```
sum_sq = sum((y-y2).^2)
ans =
     573
```

If we drew another line through the points, we could compute the sum of squares that corresponds to the new line. Of the two lines, the better fit is provided by the line with the smaller sum of squared distances. MATLAB uses techniques from calculus to minimize the sum of squared distances and arrive at the best-fit line, or the **least-squares solution**. The MATLAB commands for doing this are described in Section 8.2.3. Figure 8.7 shows the best fit results of a linear regression analysis for our data. The corresponding sum of squares is 356.8190.

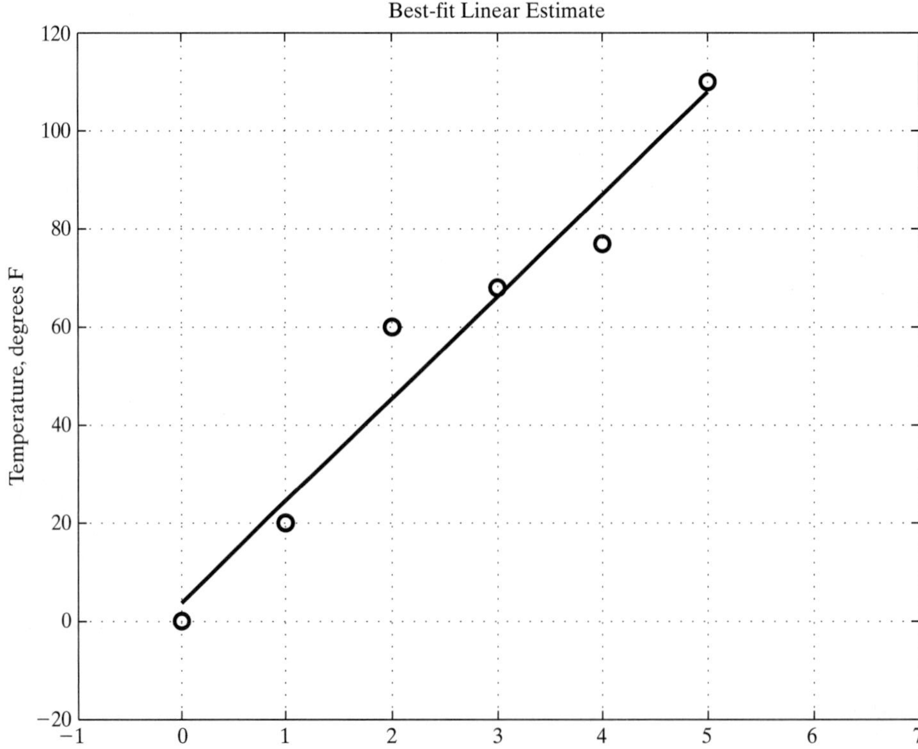

Figure 8.7
The best linear fit to a set of data points.

We call it linear regression when we derive the equation of a straight line, but more generally it is called **polynomial regression**. The linear equation used to model the data is a first-order polynomial.

8.2.2 Polynomial Regression

Linear regression is a special case of the polynomial regression technique. Recall that a polynomial with one variable can be written by using the following formula:

$$f(x) = a_0x^n + a_1x^{n-1} + a_2x^{n-2} + \cdots + a_{n-1}x + a_n$$

The **degree of a polynomial** is equal to the largest value used as an exponent. Therefore, the general form of a cubic (or third order) polynomial is

$$g(x) = a_0x^3 + a_1x^2 + a_2 + a_3$$

Note that a linear equation is also a polynomial of degree one.

In Figure 8.8, we plot the original set of data points that we used in the linear regression example, along with plots of the best-fit polynomials with degrees two through five. Note that, as the degree of the polynomial increases, the number of points that fall on the curve also increases. If a set of $n + 1$ points is used to determine an *nth* degree polynomial, all the points will fall on the polynomial.

8.2.3 polyfit and polyval Functions

The MATLAB function for computing the best fit to a set of data with a polynomial is **`polyfit`**. This function has three arguments: the *x* coordinates of the data points, the *y* coordinates of the data points, and the degree *n* of the polynomial.

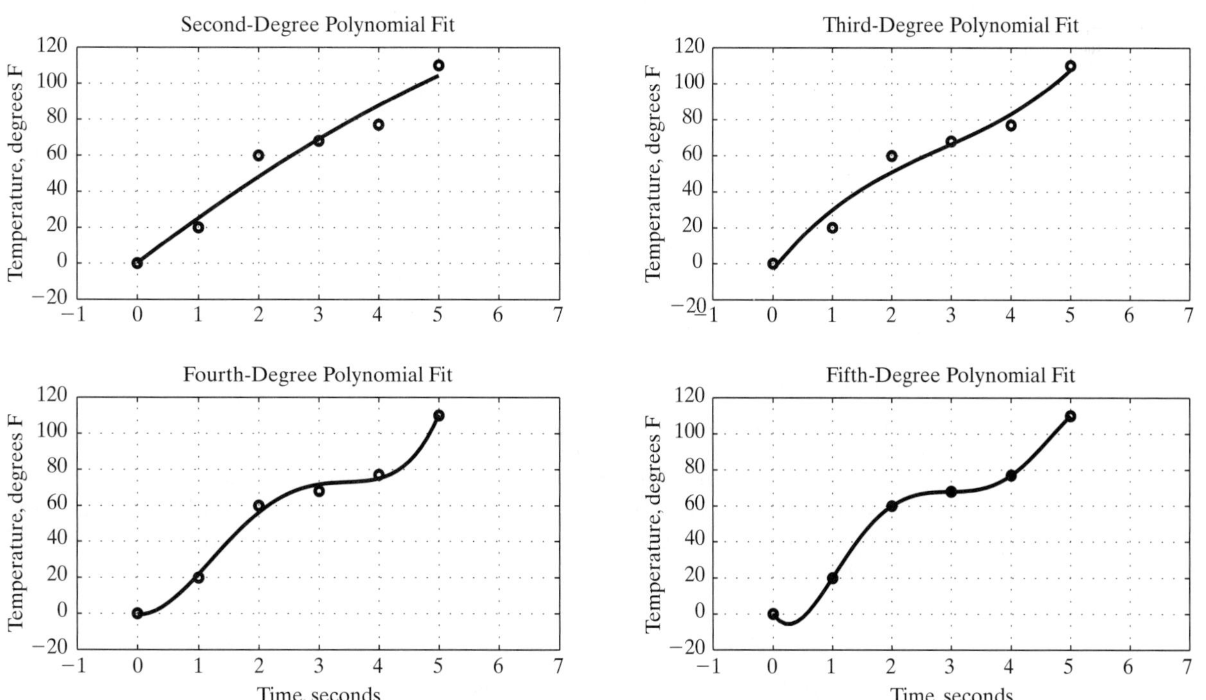

Figure 8.8
Polynomial fits.

The function returns the coefficients, in descending powers of x, of the *nth* degree polynomial used to model the data. For example, using the data

```
x = 0:5;
y = [0,20,60,68,77,110];
polyfit(x,y,1)
ans =
     20.8286   3.7619
```

So, the first-order polynomial that best fits our data is

$$f(x) = 20.8286x + 3.7619$$

Similarly, we can find other polynomials to fit the data by specifying a higher order in the **polyfit** equation. Thus,

```
polyfit(x,y,4)
ans =
       1.5625   -14.5231   38.6736   -3.4511   -0.3770
```

which corresponds to a fourth-order polynomial:

$$f(x) = 1.5625x^4 - 14.5231x^3 + 38.6736x^2 - 3.4511x - 0.3770$$

We could use these coefficients to create equations to calculate new values of y, for example,

```
y_first_order_fit = 20.8286*x + 3.7619;
y_fourth_order_fit = 1.5625*x.^4 - 14.5231*x.^3 ...
      + 38.6736*x.^2 - 3.4511*x - 0.3770;
```

or we could use the function **polyval** provided by MATLAB to accomplish the same thing.

The **polyval** function is used to evaluate a polynomial at a set of data points. The first argument of the **polyval** function is a vector containing the coefficients of the polynomial (in an order corresponding to decreasing powers of **x**), and the second argument is the vector of **x** values for which we want to calculate corresponding **y** values.

Fortunately, the **polyfit** function can provide us the first input:

```
coef = polyfit(x,y,1);
y_first_order_fit = polyval(coef,x);
```

These two lines of code could be shortened to one line by nesting functions:

```
y_first_order_fit = polyval(polyfit(x,y,1),x);
```

We can use our new understanding of the **polyfit** and **polyval** functions to write a program to create the plots in Figure 8.8. Note that we did not include the statements that would need to be added to specify the axis, the title, the x-label, and the y-label:

```
new_x = 0:0.1:5;
y2 = polyval(polyfit(x,y,2),new_x);
y3 = polyval(polyfit(x,y,3),new_x);
y4 = polyval(polyfit(x,y,4),new_x);
y5 = polyval(polyfit(x,y,5),new_x);
```

```
subplot(2,2,1),plot(x,y,'o',new_x,y2),grid,
subplot(2,2,2),plot(x,y,'o',new_x,y3),grid,
subplot(2,2,3),plot(x,y,'o',new_x,y4),grid,
subplot(2,2,4),plot(x,y,'o',new_x,y5),grid
```

The two new functions discussed in this section are summarized as follows:

`polyfit(x,y,n)` Returns a vector of $n + 1$ coefficients that represents the best-fit polynomial of degree n for the x and y coordinates provided. The coefficient order corresponds to decreasing powers of x.

`polyval(coef,x)` Returns a vector of polynomial values f(x) that correspond to the **x** vector values. The order of the coefficients corresponds to decreasing powers of x.

EXAMPLE 8.3

WATER IN A CULVERT

Determining how much water will flow through a culvert is not as easy as it might first seem (see Figure 8.9). The channel could have a nonuniform shape, obstructions might influence the flow, friction is important, and so on. A numerical approach allows us to fold all of those concerns into a model of how the water actually behaves.

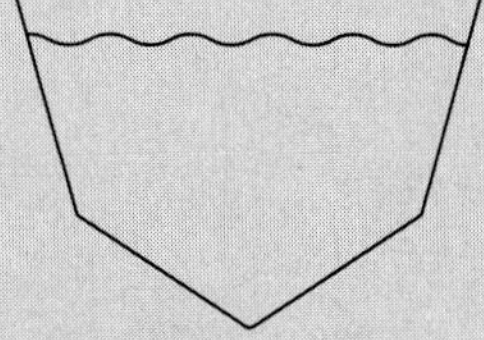

Figure 8.9
Cross section of a culvert.

Consider the data collected from an actual culvert that is contained in Table 8.4. Compute a best-fit linear, quadratic, and cubic fit for the data, and plot them on the same graph. Which model best represents the data? (Linear is first order, quadratic is second order, and cubic is third order.)

Table 8.4 Flow Measured in a Culvert

Height (ft)	Flow (ft^3/s)
0	0
1.7	2.6
1.95	3.6
2.60	4.03
2.92	6.45
4.04	11.22
5.24	30.61

SOLUTION

1. Problem Statement

Perform a polynomial regression on the data, plot the results, and determine which order best represents the data.

2. Input/Output Description

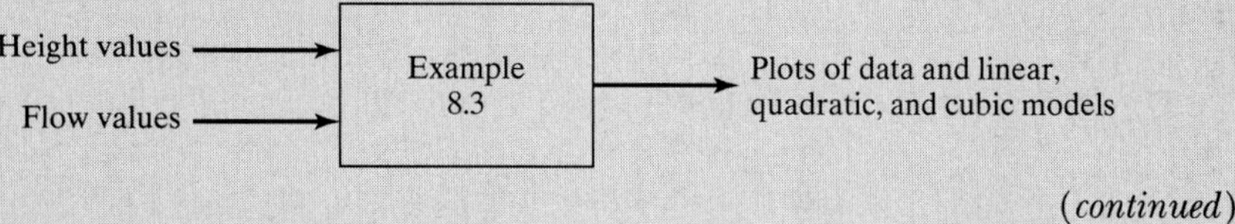

(continued)

3. Hand Example

The plot in Figure 8.10 contains a graph of the data points with a linear approximation with slope 2.0. Note that the approximation starts at zero because no water should be flowing if the height of the water in the culvert is zero. However, this model is not very accurate as the water height goes above 4 ft.

4. MATLAB Solution

We create the MATLAB solution in an M-file in order to be able to modify and rerun it.

```
%-----------------------------------------------------------
%  Example 8_3 - This program developes a model
%  for the flow of water in a culvert.
%
clear, clc
%
height = [0, 1.7, 1.95, 2.6, 2.92, 4.04, 5.24];
flow = [0, 2.6, 3.6, 4.03, 6.45, 11.22, 30.61];
%
new_height = 0:0.1:6;
newflow1 = polyval(polyfit(height,flow,1),new_height);
```

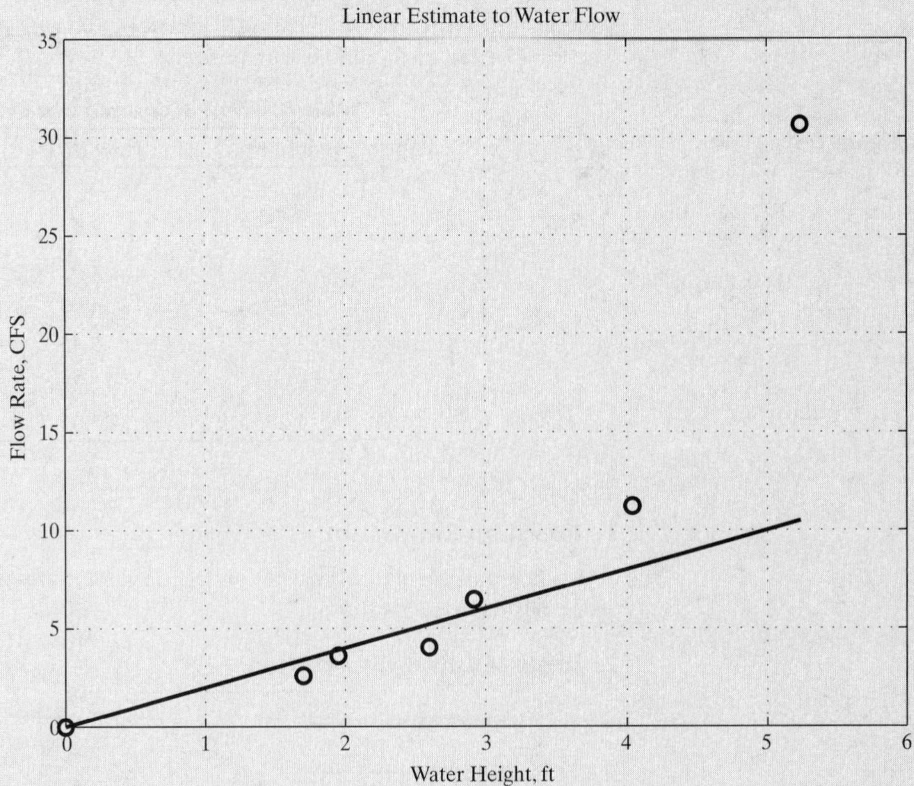

Figure 8.10
A linear fit to the water flow.

```
newflow2 = polyval(polyfit(height,flow,2),new_height);
newflow3 = polyval(polyfit(height,flow,3),new_height);
%
plot(height,flow,'o',new_height,newflow1,'--',
   new_height,newflow2,'.-',new_height,newflow3),
   title('Fit of Water Flow'),
   xlabel('Water Height, ft'),
   ylabel('Flow Rate, CFS'), grid,
   legend('Data','Linear Fit','Quadratic Fit','Cubic Fit')
%-----------------------------------------------------------
```

5. Testing

This program generates the plot shown in Figure 8.11. The question of what line best represents the data is difficult to answer. The higher order polynomial approximation will follow the data points better, but it does not necessarily represent reality better.

The linear fit predicts that the water flow rate will be approximately -5 cubic feet/s at a height of zero, which does not match reality. The quadratic fit goes back up after a minimum at a height of approximately 1.5 m—again, a result inconsistent with reality. The cubic (third-order) fit follows the points the best, and is probably the best polynomial fit.

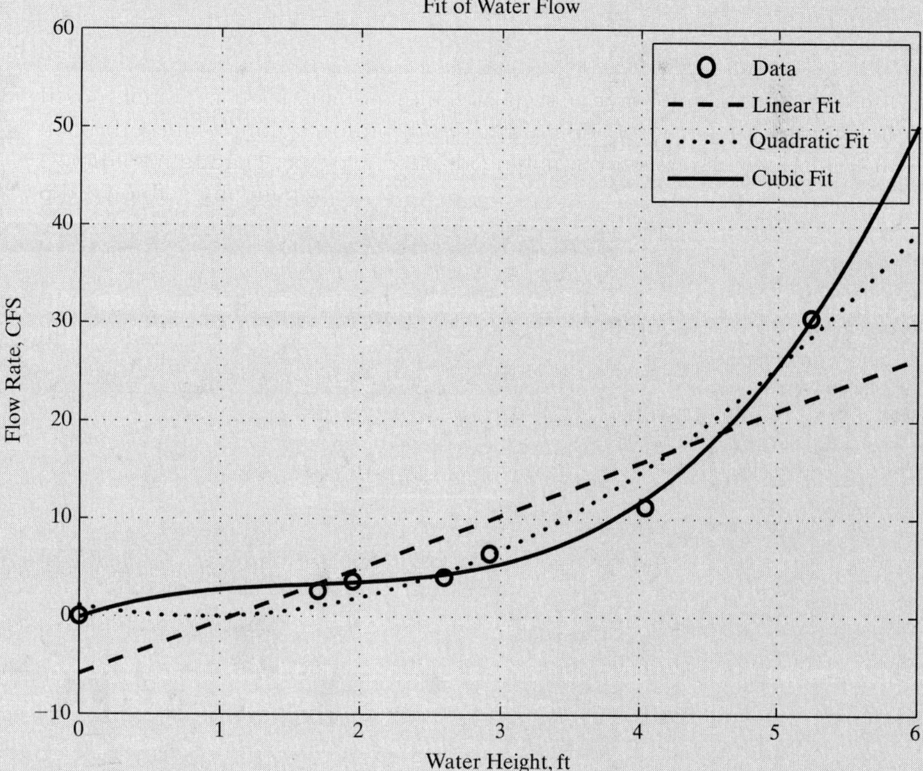

Figure 8.11
Different curve-fitting approaches.

8.3 USING THE INTERACTIVE FITTING TOOL

The interactive plotting tool allows you to annotate your plots without using the command window. The software also includes basic curve fitting, more complicated curve fitting, and statistics tools.

To access the **interactive fitting tools**, first create a plot:

```
x = 0:5;
y = [0,20,60,68,77,110];
plot(x,y,'o'),
    axis([-1,7,-20,120])
```

Maximize the plot so that it is a full-screen plot. Then activate the curve fitting tools by selecting **Tools → Basic Fitting** from the menu bar on the figure. The **Basic Fitting window** will open on top of the plot as shown in Figure 8.12. Check **linear** and **cubic** in this box, along with **show equations** and **plot residuals** to generate the plots shown in Figure 8.13. The top plot contains the linear and cubic models,

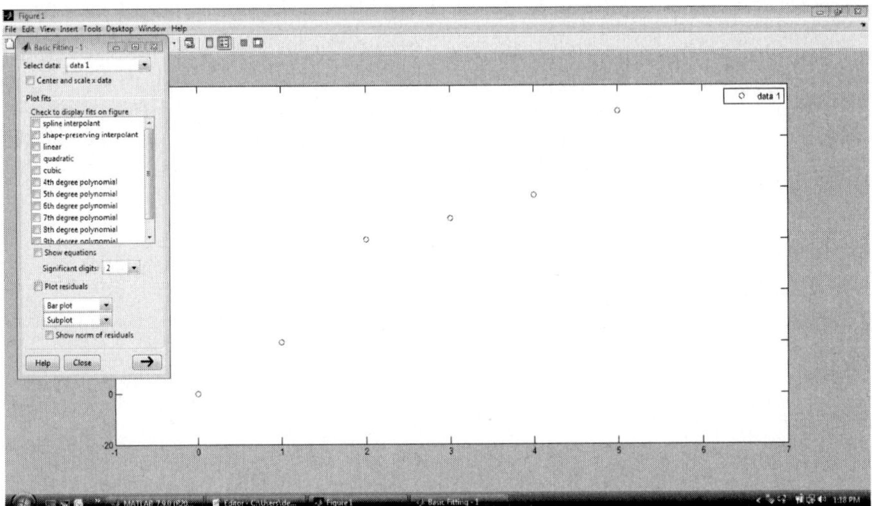

Figure 8.12
Interactive basic fitting window.

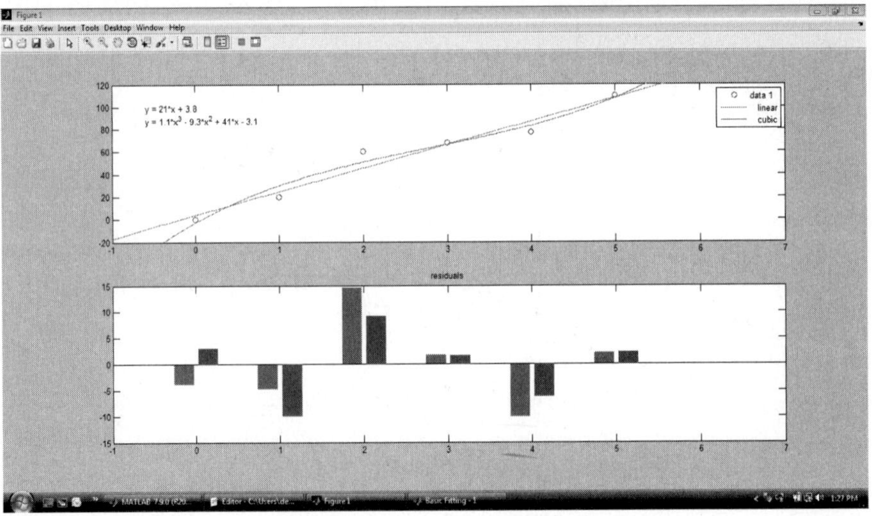

Figure 8.13
Plot generated using the basic fitting window.

along with the corresponding equations; the bottom plot shows how far each data point is from the calculated line. (Minimize the Basic Fitting window to see the complete plots.)

In the lower right-hand corner of the Basic Fitting window is an **arrow button**. Selecting that button twice opens the rest of the Basic Fitting window (Figure 8.14). The center panel of the window shows the results of the curve fit and offers the option of saving those results into the workspace. The right-hand panel allows you to select *x* values and calculate *y* values based on the equation displayed in the center panel.

In addition to the Basic Fitting window, you can also access the **Data Statistics window** (Figure 8.15) by selecting **Tools → Data Statistics** from the menu bar. This window allows you to calculate statistical functions interactively, such as mean and standard deviation, based on the data in the figure, and allows you to save the results to the workspace.

Figure 8.14
Basic Fitting window.

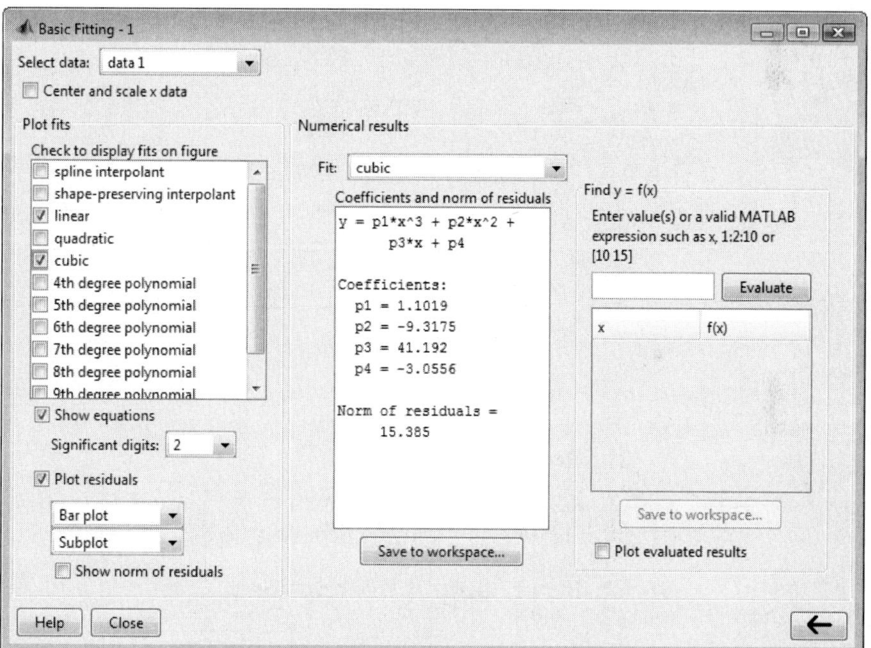

Figure 8.15
Data Statistics window.

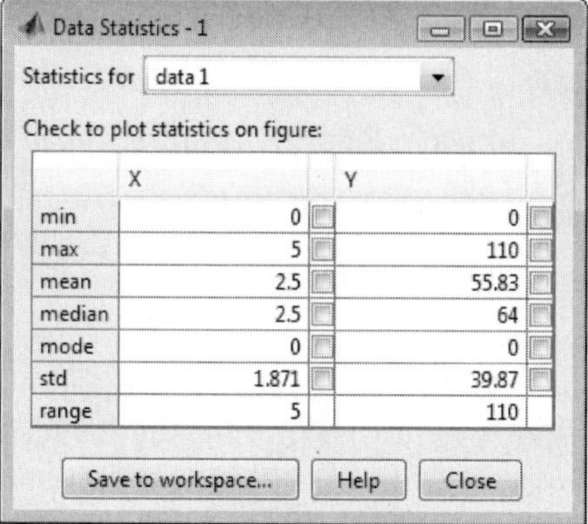

	X	Y
min	0	0
max	5	110
mean	2.5	55.83
median	2.5	64
mode	0	0
std	1.871	39.87
range	5	110

EXAMPLE 8.4

POPULATION MODELS

The population of the earth is expanding rapidly, as is the population of the United States. Table 8.5 contains the population of the U.S. Census from the years 1900 through 2000 for each decade. Use the interactive fitting tool to compare a second-degree fit to a fifth-degree fit.

Table 8.5 U.S. Census Population Data

Year	Population (millions)
1900	75.995
1910	91.972
1920	105.711
1930	123.203
1940	131.699
1950	150.697
1960	179.323
1970	203.212
1980	226.505
1990	249.633
2000	281.422

SOLUTION

1. Problem Statement

Model the population growth in the United States from 1900 through 2000 using a second-degree fit and a fifth-degree fit.

2. Input/Output Description

3. Hand Example

The plot of the data points in Figure 8.16 shows the general trend of the data.

4. MATLAB Solution

Generate an initial plot using these statements:

```
years = 1900:10:2000;
population = [75.995,91.972,105.711,123.203,131.699,...
   150.697,179.323,203.212,226.505,249.633,281.422];
plot(years,population,'o')
```

Then use the interactive tool to do the interpolations and to plot the two models.

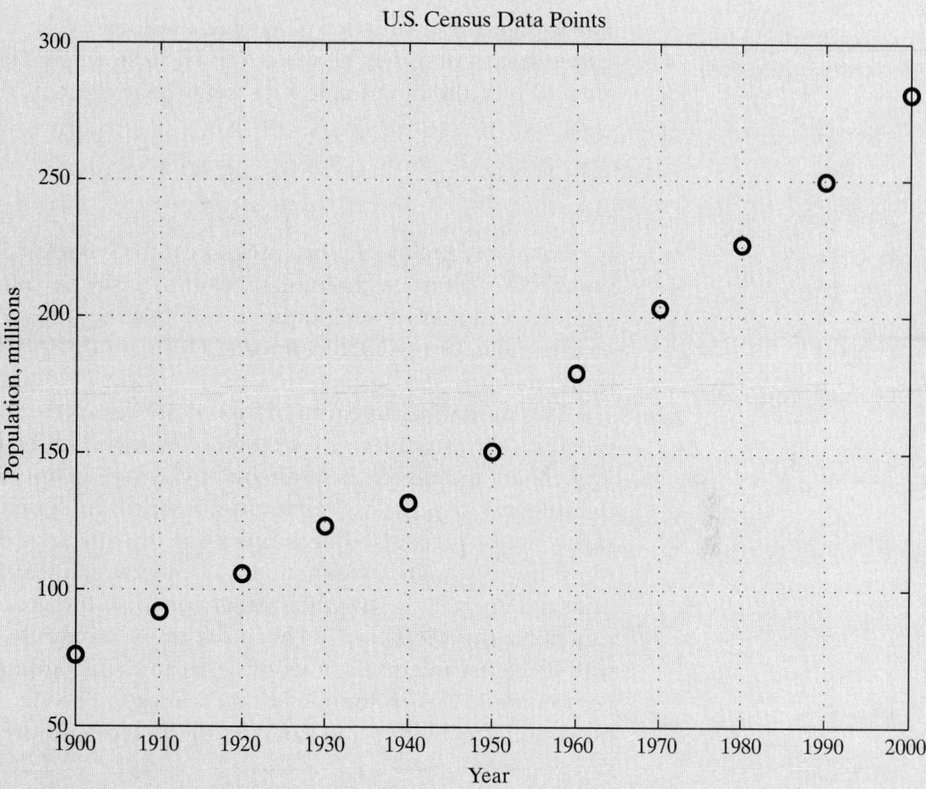

Figure 8.16
Census data points.

5. Testing

Figure 8.17 contains a plot of the two models generated with the interactive fitting tool. We also selected **Show norm of residuals** in addition to plotting the residuals. Compare the fits—they both appear to model the data adequately. The residuals norms show that the fifth-degree model is a better fit than the quadratic fit; however, the two residuals are reasonable close in value. Since both models are good, we would need to weigh the advantage of a better fit with the fifth-degree model with the simpler equation with a quadratic equation.

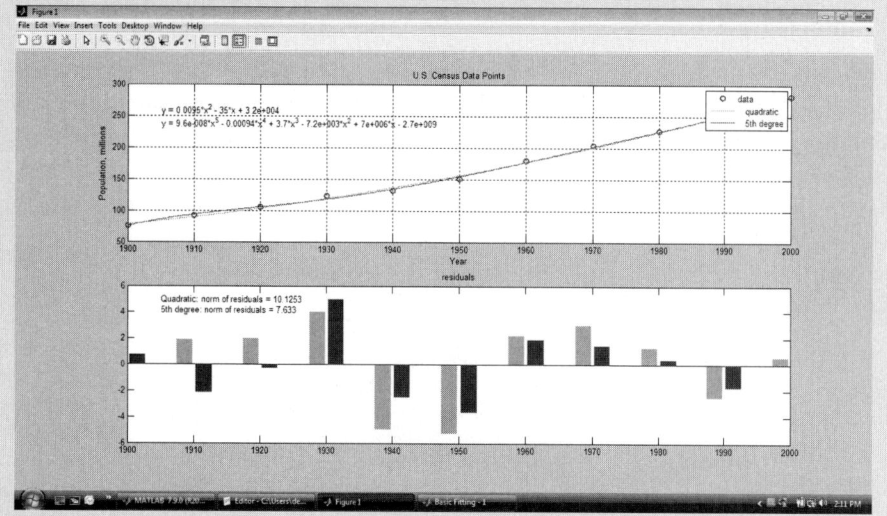

Figure 8.17
Two models for U.S. Census data.

8.4 NUMERICAL INTEGRATION

The **integral** of a function $f(x)$ over the interval $[a, b]$ is defined to be the area under the curve of $f(x)$ between a and b, as shown in Figure 8.18. If the value of this integral is K, the notation to represent the integral of $f(x)$ between a and b is

$$K = \int_a^b f(x)\,dx$$

For many functions, this integral can be computed analytically. However, for a number of functions, the integral cannot easily be computed analytically and thus requires a numerical technique to estimate its value. The numerical evaluation of an integral is also called **quadrature**, a term that comes from an ancient geometrical problem.

The **numerical integration** techniques estimate the function $f(x)$ by another function $g(x)$, where $g(x)$ is chosen so that we can easily compute the area under $g(x)$. Then, the better the estimate of $g(x)$ to $f(x)$, the better will be the estimate of the integral of $f(x)$. Two common numerical integration techniques estimate $f(x)$ with a set of piecewise linear functions or with a set of piecewise parabolic functions. If we estimate the function with piecewise linear functions, we can then compute the area of the trapezoids that compose the area under the piecewise linear functions; this technique is called the **trapezoidal rule**. If we estimate the function with piecewise quadratic functions, we can then compute and add the areas of these components; this technique is called **Simpson's rule**. MATLAB uses an advanced quadrature technique and can typically include infinity for one of the end points of the integration.

The MATLAB function for performing numerical integration is called **`integral`**, and it is referenced as follows:

`integral(@function,a,b)`	Returns the area of the **`function`** between **a** and **b**, assuming that **`function`** is a MATLAB function (either a built-in function or a user-written function)

To illustrate the use of this function, assume that we want to determine the **`integral`**, of the square-root function for nonnegative values of a and b:

$$K = \int_a^b \sqrt{x}dx$$

The square-root function $f(x) = \sqrt{x}$ is plotted in Figure 8.19 for the interval $[0, 5]$; the values of the function are complex for $x < 0$. This function can be integrated analytically to yield the following for nonnegative values of a and b:

$$K = \frac{2}{3}(b^{3/2} - a^{3/2})$$

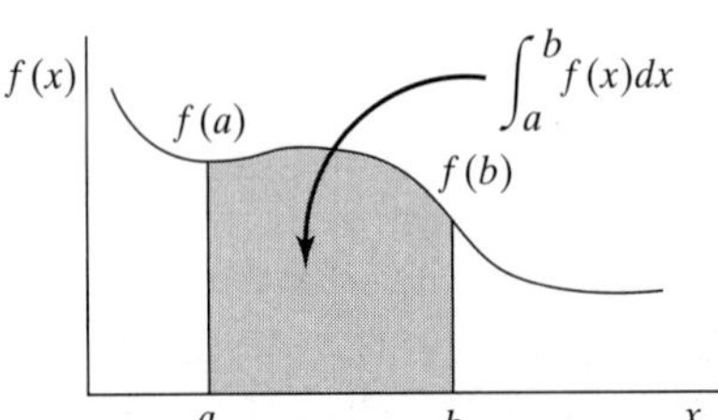

Figure 8.18
Integral of f(x) from a to b.

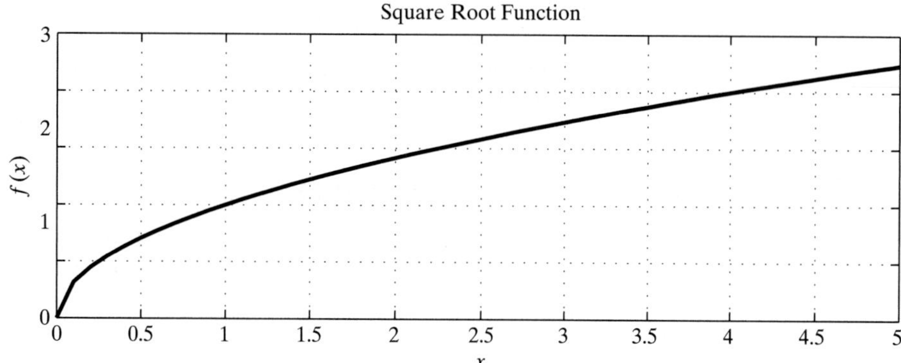

Figure 8.19
Square-root function.

The MATLAB program shown next can be used to compare the results of the **integral** function with the analytically calculated results. The script prompts the user for a specified interval:

```
%-----------------------------------------------------------
%   These statements compare the integral function
%   with the analytical results for the integration of the
%   square root of x over an interval [a,b], where a and b
%   are nonnegative.
%
a = input('Enter left endpoint (nonnegative): ');
b = input('Enter right endpoint (nonnegative): ');
%
%   k is the computed analytical result
k = (2/3)*(b^(1.5) - a^(1.5));
%
%   The following statement computes the integral
%   function from a to b.
k_est = integral(@sqrt,a,b);
%
%  Display the results
disp('Exact Integral:')
disp(k)
disp('Estimated Integral:')
disp(est_k)
%-----------------------------------------------------------
```

The following example demonstrates the script's use. Note that for these two intervals, the analytical result is the same as the numerical integration result.

```
Enter left endpoint (nonnegative): 1.5
Enter right endpoint (nonnegative): 15
Exact Integral:
   37.5051
Estimated Integral:
   37.5051
Enter left endpoint (nonnegative): .01
Enter right endpoint (nonnegative): .011
```

```
Exact Integral:
    1.0246e-04
Estimated Integral:
    1.0246e-04
```

8.5 NUMERICAL DIFFERENTIATION

The **derivative** of a function $f(x)$ is defined to be a function f' that is equal to the rate of change of $f(x)$ with respect to x. The derivative can be expressed as a ratio, with the change in $f(x)$ indicated by $df(x)$ and the change in x indicated by dx, giving

$$f'(x) = \frac{df(x)}{dx}$$

There are many physical processes for which we want to measure the rate of change of a variable. For example, velocity is the rate of change of position (as in meters per second), and acceleration is the rate of change of velocity (as in meters per second squared). It can also be shown that the integral of acceleration is velocity and that the integral of velocity is position. Hence, integration and differentiation have a special relationship, in that they can be considered to be inverses of each other: The derivative of an integral returns the original function, and the integral of a derivative returns the original function, to within a constant value.

The derivative $f'(x)$ can be described graphically as the slope of the function $f(x)$, where the slope of $f(x)$ is defined to be the slope of the tangent line to the function at the specified point. Thus, the value of $f'(x)$ at the point a is $f'(a)$ and it is equal to the slope of the tangent line at the point a, as shown in Figure 8.20.

Because the derivative of a function at a point is the slope of the tangent line at the point, a value of zero for the derivative of a function at the point x_k indicates that the line is horizontal at that point. Points with derivatives of zero are called **critical points** and can represent either a horizontal region, a local maximum, or a local minimum of the function. (The point may also be the global maximum or global minimum, as shown in Figure 8.21, but more analysis of the entire function would be needed to determine this.) If we evaluate the derivative of a function at several points in an interval and we observe that the sign of the derivative changes, then a local maximum or a local minimum occurs in the interval. The second derivative [the derivative of $f'(x)$] can be used to determine whether or not the critical points represent local maxima or local minima. More specifically, if the second derivative of a critical is positive, then the value of the function at that point is a local minimum; if the second derivative is negative, then the value of the function at the critical point is a local maximum.

Figure 8.20
Derivative of $f(x)$ at $x = a$.

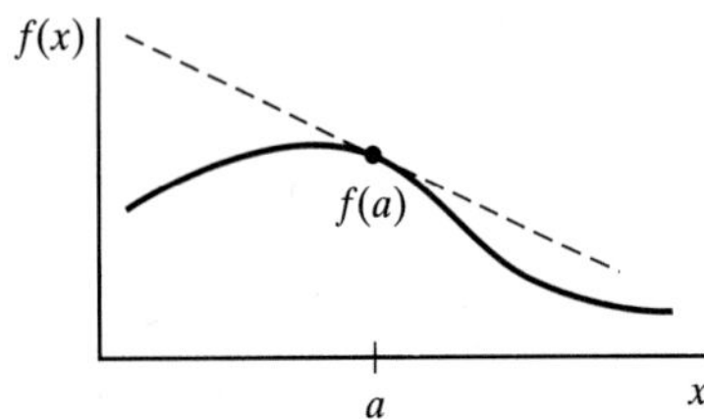

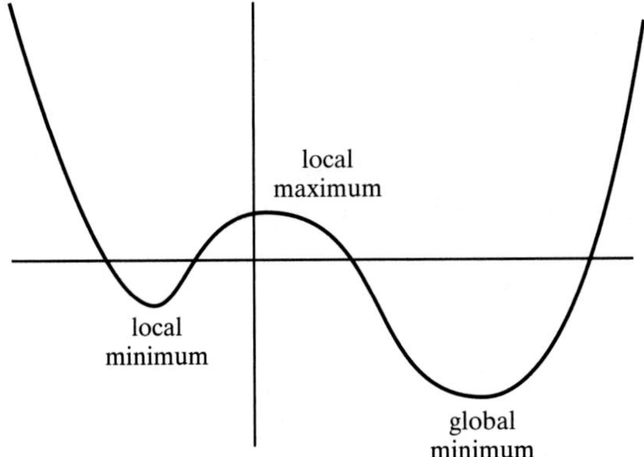

Figure 8.21
Example of function with critical points.

8.5.1 Difference Expressions

Numerical differentiation techniques estimate the derivative of a function at a point x_k by approximating the slope of the tangent line at x_k using values of the function at points near x_k. The approximation of the slope of the tangent line can be done in several ways, as shown in Figure 8.22.

Figure 8.22(a) assumes that the derivative at x_k is estimated by computing the slope of the line between $f(x_{k-1})$ and $f(x_k)$, as in

$$f'(x_k) = \frac{f(x_k) - f(x_{k-1})}{x_k - x_{k-1}}$$

This type of derivative approximation is called a **backward difference** approximation.

Figure 8.22(b) assumes that the derivative at x_k is estimated by computing the slope of the line between $f(x_k)$ and $f(x_{k+1})$, as in

$$f'(x_k) = \frac{f(x_{k+1}) - f(x_k)}{x_{k+1} - x_k}$$

This type of derivative approximation is called a **forward difference** approximation.

Figure 8.22(c) assumes that the derivative at x_k is estimated by computing the slope of the line between $f(x_{k-1})$ and $f(x_{k+1})$, as in

$$f'(x_k) = \frac{f(x_{k+1}) - f(x_{k-1})}{x_{k+1} - x_{k-1}}$$

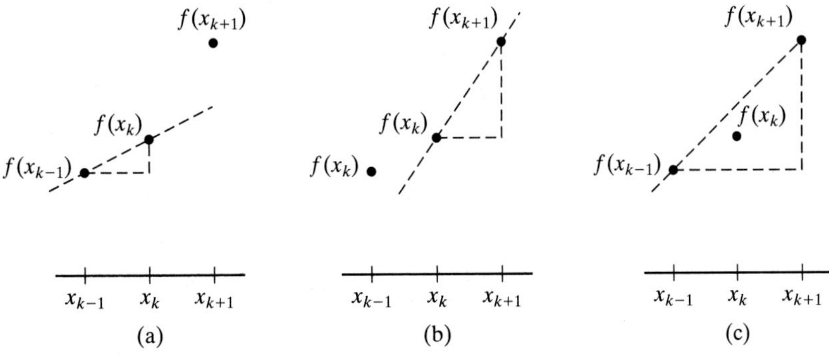

Figure 8.22
Techniques for computing $f'(x_k)$.

This type of derivative approximation is called a **central difference** approximation, and we usually assume that x_k is halfway between x_{k-1} and x_{k+1}.

The quality of all of these types of derivative computations depends on the distance between the points used to estimate the derivative; the estimate of the derivative improves as the distance between the two points decreases.

The second derivative of a function $f(x)$ is the derivative of the first derivative of the function:

$$f''(x) = \frac{df'(x)}{dx}$$

This function can be evaluated using slopes of the first derivative. Thus, if we use backward differences, we have

$$f''(x_k) = \frac{f'(x_k) - f'(x_{k-1})}{x_k - x_{k-1}}$$

Similar expressions can be derived for computing estimates of higher derivatives.

8.5.2 diff Function

The **diff** function computes differences between adjacent values in a vector, generating a new vector with one fewer value. If the **diff** function is applied to a matrix, it operates on the columns of the matrix as if each column were a vector. A second, optional argument specifies the number of times to recursively apply **diff**. Each time **diff** is applied, the length of the vector is reduced in size. A third, optional argument specifies the dimensions in which to apply the function. The forms of **diff** are summarized as follows:

`diff(X)`	For a vector `X`, diff returns `[X(2)-X(1),X(3)-X(2), ... ,X(n)-X(n-1)].`
`diff(X)`	For a matrix `X`, diff returns the matrix of column differences
`diff(X,n,dim)`	The general form of **diff** returns the nth difference function along dimension **dim** (a scalar). If n is greater than or equal to the length of **dim**, then **diff** returns an empty array.

To illustrate, we define vectors **x**, **y**, and **z** as follows:

```
x = [0,1,2,3,4,5];
y = [2,3,1,5,8,10];
z = [1,3,5;1,5,10];
```

We now illustrate several different references to the **diff** function. Go through these carefully to be sure that you understand how this function works for different inputs.

```
diff(x)
ans =
     1   1   1   1   1
diff(y)
ans =
     1  -2   4   3   2
diff(y,2)
ans =
    -3   6  -1  -1
```

```
diff(z,1,1)
ans =
       0    2    5
diff(z,1,2)
ans =
       2    2
       4    5
```

An approximate derivative *dy* can be computed by using **`diff(y)./diff(x)`**. Note that these values of *dy* are correct for both the forward difference equation and the backward difference equation. The distinction between the two methods for computing the derivative is determined by the values of the vector **`xd`**, which correspond to the derivative *dy*. If the corresponding values of **`xd`** are [1, 2, 3, 4, 5], *dy* computes a backward difference. If the corresponding values of **`xd`** are [0, 1, 2, 3, 4], *dy* computes a forward difference.

As an example, consider the function given by the following polynomial:

$$f(x) = x^5 - 3x^4 - 11x^3 + 27x^2 + 10x - 24$$

A plot of this function is shown in Figure 8.23. Recall that the zeros of the derivative correspond to the points of local minima or local maxima of a function. The function in this example does not have a global minimum or global maximum, because the function ranges from $-\infty$ to ∞. The local minima and maxima (or critical points) of this function occur at -2.3, -0.2, 1.5, and 3.4. Assume that we want to compute the derivative of this function over the interval $[-4, 5]$. We can perform this operation using the **`diff`** function, as shown in the following script, where **`df`** represents *df* and **`xd`** represents the *x* values corresponding to the derivative:

```
%   Evaluate f(x) and f'(x).
%
x = -4:0.1:5;
f = x.^5 - 3*x.^4 - 11*x.^3 + 27*x.^2 + 10*x - 24;
df = diff(f)./diff(x);
xd = x(2:length(x));
plot(xd,df);
```

This plot is shown in Figure 8.24.

Using values of **`df`** from the previous calculations, we can use the **`find`** function to determine the indices *k* of the locations in the product of consecutive values that give a negative value. These are points where the slope changes sign, and thus

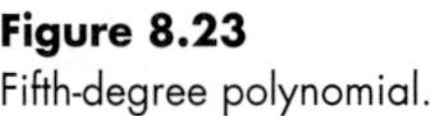
Figure 8.23
Fifth-degree polynomial.

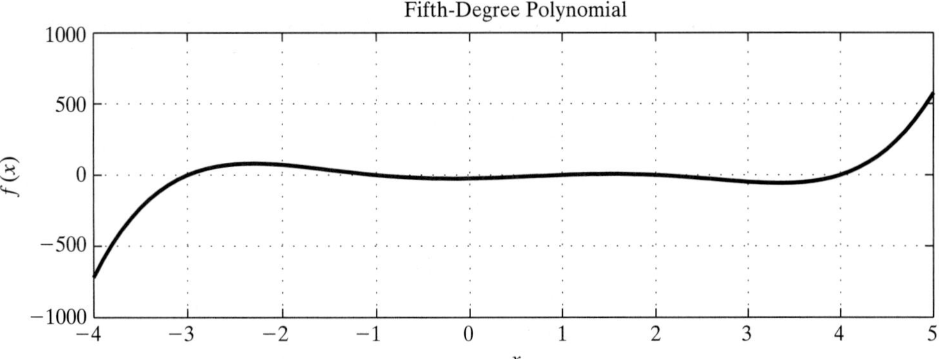

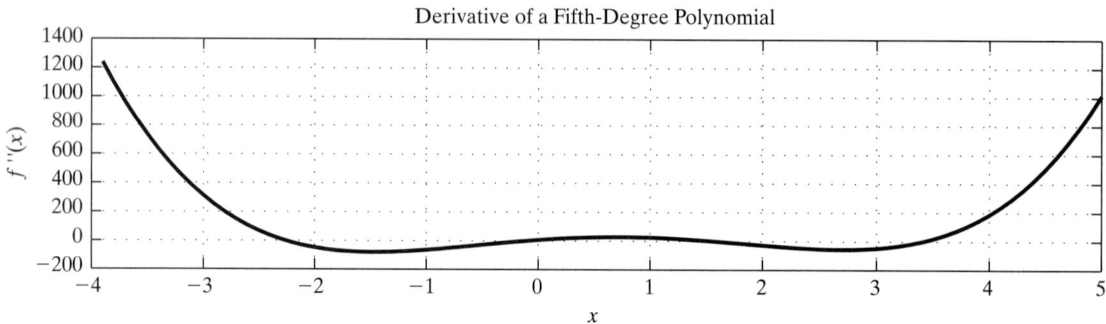

Figure 8.24
Derivative of a fifth-degree polynomial.

crosses zero. These indices are then used with the vector **xd** to print the approximation to the locations of the critical points:

```
%  Find locations of critical points of f'(x).
%
product = df(1:length(df) - 1).*df(2:length(df));
critical = xd(find(product<0))
critical =
          -2.3000   -0.2000   1.5000   3.4000
```

In the example discussed in this section, we assumed that we had the equation of the function to be differentiated, and thus we could generate points of the function. In many engineering problems, the data to be differentiated are collected from experiments. Thus, we cannot choose the points to be close together to get a more accurate measure of the derivative. In these cases, it might be a good solution to use alternative techniques that allow us to determine an equation for a polynomial that fits a set of data and then compute points from the equation to use in computing values of the derivative.

SUMMARY

In this chapter, we explained the difference between interpolation and least-squares curve fitting. Two types of interpolation were presented: linear interpolation and cubic-spline interpolation. After presenting the MATLAB commands for performing these types of interpolations, we then turned to least-squared curve fitting using polynomials. This discussion explained how to determine the best fit to a set of data using a polynomial with a specified degree and then how to use the best-fit polynomial to generate new values of the function. Use of the interactive basic fitting window was also described to perform these same functions. Techniques for numerical integration and numerical differentiation were also presented in this chapter.

MATLAB Summary

This MATLAB summary lists and briefly describes all of the commands and functions that were defined in this chapter:

Commands and Functions	
`diff`	computes the differences between adjacent values
`integral`	computes the integral under a curve (Simpson)
`interp1`	computes linear and cubic interpolation
`polyfit`	computes a least-squares polynomial
`polyval`	evaluates a polynomial

KEY TERMS

backward difference
basic fitting window
best fit
central difference
critical points
cubic spline
data statistics window
degree of a polynomial
derivative
forward difference
integral
interactive fitting tools
least-squares solution
linear interpolation
linear regression
numerical differentiation
numerical integration
polynomial regression
quadrature
residual sum
Simpson's rule
trapezoidal rule

PROBLEMS

INTERPOLATION

Generate $f(x) = x^2$ for $x = [-3, -1, 0, 2, 5, 6]$.

1. Compute and plot the linear and cubic-spline interpolation of the data points over the range $[-3{:}0.5{:}6]$.
2. Compute the value of $f(4)$ using linear interpolation and cubic-spline interpolation. What are the respective errors when the answer is compared with the actual value of $f(4)$?

CYLINDER HEAD TEMPERATURES

Assume that the set of temperature measurements in Table 8.6 is taken from the cylinder head in a new engine that is being tested for possible use in a race car.

Table 8.6 Cylinder Head Temperatures

Time (s)	Temperature (°F)
0.0	0.0
1.0	20.0
2.0	60.0
3.0	68.0
4.0	77.0
5.0	110.0

3. Compare plots of these data, assuming linear interpolation and assuming cubic-spline interpolation for values between the data points, using time values from 0 to 5 in increments of 0.1 s.
4. Using the data from Problem 3, find the time value for which there is the largest difference between its linear-interpolated temperature and its cubic-interpolated temperature.

EXPANDED CYLINDER HEAD DATA

Assume that we measure temperatures at three points around the cylinder head in the engine instead of at just one point. Table 8.7 contains this expanded set of data.

Table 8.7 Expanded Cylinder Head Temperatures, °F

Time (s)	Temp1	Temp2	Temp3
0.0	0.0	0.0	0.0
1.0	20.0	25.0	52.0
2.0	60.0	62.0	90.0
3.0	68.0	67.0	91.0
4.0	77.0	82.0	93.0
5.0	110.0	103.0	96.0

5. Assume that these data have been stored in a matrix with six rows and four columns. Determine interpolated values of temperature at the three points in the engine at 2.6 seconds, using linear interpolation.
6. Using the information from Problem 5, determine the time that the temperature reached 75 degrees at each of the three points in the cylinder head.

SPACECRAFT ACCELEROMETER

The guidance and control system for a spacecraft often uses a sensor called an accelerometer, which is an electromechanical device that produces an output voltage proportional to the applied acceleration. Assume that an experiment has yielded the set of data shown in Table 8.8.

Table 8.8 Spacecraft Accelerometer Data

Acceleration	Voltage
–4	0.593
–2	0.436
0	0.061
2	0.425
4	0.980
6	1.213
8	1.646
10	2.158

7. Determine the linear equation that best fits this set of data. Plot the data points and the linear equation.

8. Determine the sum of the squares of the distances of these points from the line of best fit determined in Problem 7.
9. Compare the error sum from Problem 8 with the same error sum computed from the best quadratic fit. What do these sums tell you about the two models for the data?

TANGENT FUNCTION

Compute tan(*x*) for *x* = [-1:0.05:1].

10. Compute the best-fit polynomial of order four that approximates tan(*x*). Plot tan(*x*) and the generated polynomial on the same graph. What is the sum of squared errors of the polynomial approximation for the data points in *x*?
11. Compute tan(*x*) for *x* = [-2:0.05:2]. Using the polynomial generated in Problem 10, compute values of *y* from −2 to 2, corresponding to the **x** vector just defined. Plot tan(*x*) and the values generated from the polynomial on the same graph. Why aren't they the same shape?

SOUNDING ROCKET TRAJECTORY

The data set in Table 8.9 represents the time and altitude values for a sounding rocket that is performing high-altitude atmospheric research on the ionosphere.

Table 8.9 Sounding Rocket Trajectory Data

Time, s	Altitude, m
0	60
10	2,926
20	10,170
30	21,486
40	33,835
50	45,251
60	55,634
70	65,038
80	73,461
90	80,905
100	87,368
110	92,852
120	97,355
130	100,878
140	103,422
150	104,986
160	106,193
170	110,246
180	119,626
190	136,106
200	162,095
210	199,506

(continued)

Time, s	Altitude, m
220	238,775
230	277,065
240	314,375
250	350,704

12. Determine an equation that represents the data, using the interactive curve fitting tools.
13. Plot the altitude data. The velocity function is the derivative of the altitude function. Using numerical differentiation, compute the velocity values from these data, using a backward difference. Plot the velocity data. (Note that the rocket is a two-stage rocket.)
14. The acceleration function is the derivative of the velocity function. Using the velocity data determined from Problem 13, compute the acceleration data, using backward difference. Plot the acceleration data.

SIMPLE ROOT FINDING

Even though MATLAB makes it easy to find the roots of a function, sometimes all that is needed is a quick estimate. This can be done by plotting a function and zooming in very close to see where the function equals zero. Since MATLAB draws straight lines between data points in a plot, it is good to draw circles or stars at each data point, in addition to the straight lines connecting the points.

15. Plot the following function, and zoom in to find the roots:

```
n = 5;
x = linspace(0,2*pi,n);
y = x.*sin(x) + cos(1/2*x).^2 - 1./(x-7);
plot(x,y,'-o')
```

16. Increase the value of **n** to increase the accuracy of the estimate in Problem 15.

DERIVATIVES AND INTEGRALS

Consider the data points in the following two vectors:

```
X = [0.1,0.3,5,6,23,24];
Y = [2.8,2.6,18.1,26.8,486.1,530];
```

17. Determine the best-fit polynomial of order 2 for the data. Calculate the sum of squares for your results. Plot the best-fit polynomial for the six data points.
18. Generate a new **X** containing 250 uniform data points in increments of 0.1 from [0.1, 25.0]. Using the best-fit polynomial coefficients from the previous problem, generate a new **Y** containing 250 data points. Plot the results.
19. Compute an estimate of the derivative using the new **X** and the new **Y** generated in the previous problem. Compute the coefficients of the derivative. Plot the derivative using differences along with the points computed from using the equation of the derivative (determined from the coefficients).
20. Let the function f be defined by the following equation:

$$f(x) = 4e^{-x}$$

Plot this function over the interval [0, 1]. Use numerical integration techniques to estimate the integral of $f(x)$ over [0, 0.5], and over [0, 1].

Index